Two years ago, a teacher in Spain used an AI writing assistant to cut her grading time in half.

AI is no longer a tool for experts — it's becoming a part of everyday life for all of us.

A doctor in India used a diagnostic model to spot a rare disease early.

A student in Colombia created art that won a global competition — with the help of a generative tool.

These moments aren't from the future — they're happening right now.

Artificial intelligence is no longer science fiction — it's the invisible engine shaping how we learn, create, heal, and work. You already interact with it every day: when you ask a smart assistant a question, stream music, receive a medical result, or scroll through social media.

But what is AI, really? How does it think, decide, and create? And — most importantly — how can we live and work wisely with it instead of being overwhelmed by it?

Mastering AI for Everyone is your clear, human-centered guide to understanding and using AI confidently. In the pages ahead, you'll discover what AI can do, what it can't, and how to make it work for you—not the other way around.

By the end of this book, you'll not only understand artificial intelligence — you'll know how to think with it, question it, and apply it to your own goals.

MASTERING AI FOR EVERYONE

A Clear, Practical Guide to Understanding Artificial Intelligence and Using It in Everyday Life

ANGEL VALE

First Edition

ISBN: 978-1-971032-00-9

Mastering AI
PUBLISHING

Disclaimer

This book is intended for **educational and informational purposes** only. It explores the concepts, applications, benefits, and challenges of artificial intelligence (AI) based on knowledge and tools available at the time of writing.

The AI technologies, platforms, and resources referenced throughout the book — including tools such as ChatGPT, Midjourney, Notion AI, and others — are **mentioned as illustrative examples** and do not constitute endorsements, sponsorships, or affiliations with any specific companies or products.

While every effort has been made to ensure the **accuracy, clarity, and reliability** of the information, readers should be aware that AI is a **rapidly evolving field**. Features, capabilities, usage policies, and ethical considerations may change significantly over time. Readers are encouraged to verify current details, and exercise discretion before relying on any tool or service for personal or professional use.

The use of AI tools may also be subject to **local regulations, data privacy laws, and compliance frameworks**, which can vary depending on region, industry, and application. It is the reader's responsibility to ensure they adhere to any applicable legal or ethical requirements when engaging with AI technologies.

AI systems mentioned in this book may display unexpected behaviors, inaccuracies, or limitations. Their outputs should always be evaluated with **critical thinking and human oversight**, especially in sensitive domains like healthcare, law, education, and finance.

This book was created with the assistance of **AI tools** to support structure, clarity, and visual elements. However, all ideas, recommendations, and content have been **carefully curated and reviewed** by human authors to ensure consistency, accuracy, and contextual relevance.

Ultimately, the goal of this book is to empower readers to **better understand and responsibly navigate AI** — not to offer technical, legal, or business advice.

Author's Note

Mastering AI began with a simple belief:

everyone — not just engineers, researchers, or big tech companies — deserves to understand artificial intelligence clearly and confidently.

This book is part of that mission.

From the start, the goal wasn't just to write a guide. It was to build an ecosystem — a collection of books, courses, tools, and resources designed to help people navigate the new AI era without fear, confusion, or hype. And while I wrote this book under the name Angel Vale, it reflects the work of a broader team: researchers, editors, educators, and creators who share a commitment to making AI accessible and responsible for all.

We created this book because AI is now part of everyday life — shaping how we learn, work, create, decide, and connect. But technology alone isn't enough. What matters is how people use it, question it, and shape it. That's why clarity, transparency, and practical guidance were at the heart of every chapter.

This book was created to help everyday people understand AI without needing a technical background.

Thank you for being curious, critical, and open-minded.

Thank you for learning with us.

And thank you for being part of the larger journey toward an informed, thoughtful, and human-centered AI future.

— **Angel Vale**

on behalf of the Mastering AI team

About the Author

Angel Vale writes as part of the Mastering AI team, a collective dedicated to making artificial intelligence clear, accessible, and genuinely useful for everyday people. With a background that blends technology, communication, and practical problem-solving, Angel focuses on explaining complex ideas in a grounded, human way.

This book is part of a broader mission: to help readers understand AI with confidence, without hype or confusion.

TABLE OF CONTENTS

Appendix

How to Use This Book

This book is meant to be practical, flexible, and easy to navigate — a clear guide to understanding artificial intelligence whether you read it cover-to-cover or jump to the topics that matter most to you. Here's how to get the most out of it:

1. You can read it straight through — or in any order.

Chapters build naturally from foundations → real-world uses → deeper concepts → ethics and the future.

But each chapter is written to stand on its own. Explore in whatever sequence fits your goals.

2. Use the structure to guide your learning.

Each chapter is organized the same way:

- Clear explanations of concepts
- Real examples from everyday life
- Brief takeaways highlighting what matters
- A short summary to reinforce the main ideas

If you skim, the takeaways and summaries give you the essentials quickly.

3. Don't worry about technical details.

Whenever a new term appears, it's explained in plain language — so you can learn AI without a technical background..

More formal definitions are collected in the Glossary at the end for quick reference.

4. Apply what you learn as you go.

Throughout the book, you'll find:

- prompts to reflect on your own work, habits, and tools
- real use cases you can try immediately

- everyday-life examples that show how AI actually works around you
- examples designed to spark ideas for your personal or professional life

If something resonates, pause and test it out.

5. Use the Appendix as your reference toolkit.

The back of the book includes:

- a Glossary of Terms
- curated Resources for Further Learning
- links and tools for continuing your AI journey

These are practical resources you can return to whenever you start exploring AI tools more deeply.

6. Read actively — not passively.

You'll get more out of the book if you:

- highlight concepts that matter to you
- jot down ideas for how AI could help in your daily routine
- flag questions you want to explore later

AI is changing fast; thinking as you read prepares you to adapt confidently.

7. This book is a starting point, not an ending.

AI evolves quickly. The purpose of this book is to give you a foundation strong enough to keep learning — with clarity instead of confusion.

Introduction

Artificial intelligence is no longer the stuff of science fiction. It's here — in our pockets, in our workplaces, even helping diagnose diseases and generate music. **But for most people, understanding AI still feels confusing — and that's exactly what this book helps solve.** And yet, for many people, AI still feels like a black box: powerful, mysterious, and maybe even a little intimidating.

I wrote this book because I believe AI shouldn't just be understood by engineers, tech giants, or data scientists. It should be understood — and used — by you.

What began as a personal learning journey quickly became something bigger. As someone who's always been passionate about technology, I immersed myself in AI — through courses, hands-on experiments, and real-world consulting work. I saw how AI could unlock productivity, creativity, and entirely new ways of thinking. But I also saw something else: a growing gap between what AI is capable of and what most people actually understand or use.

That's why I created Mastering AI — not just a book series, but a broader effort to make artificial intelligence understandable, accessible, and practical for everyone.

It now includes over 30 books in English and Spanish, a growing online resource hub at masteringai.ai, and a simple mission: **to make artificial intelligence accessible, practical, and easy to understand for everyone — so people can thrive, not just adapt, in an AI-powered world.**

This first book, Mastering AI for Everyone, is the foundation. It's designed to demystify artificial intelligence and show how it's already reshaping the world around you—often invisibly.

Whether you're a business owner looking to improve efficiency, a teacher curious about new tools, or simply someone trying to keep up with the pace of change, this book is your roadmap.

Inside, you'll discover:

- What AI really is — and what it isn't
- How it works beneath the surface (in plain language)
- Where it shows up in your daily life — from your phone to your car

- How to use AI tools to save time, work smarter, and spark creativity
- The risks, biases, and big ethical questions we all need to face together

This isn't a technical manual, and you won't need a background in math or programming to follow along. Think of it as a guided tour—from the core concepts of how AI functions to the real ways it's changing jobs, decisions, education, creativity, and privacy.

By the end of this book, you won't just understand how AI works — you'll be able to use it with purpose, insight, and confidence.

Artificial intelligence is here. The question is no longer if you'll use it — it's how.

And this is just the beginning. As Mastering AI grows, we'll continue building guides, resources, and practical tools to help you stay ahead — and use AI wisely.

Mastering AI is an independent project dedicated to helping people understand, use, and navigate artificial intelligence with clarity and responsibility.

We combine research, practical guidance, and real-world examples to make AI accessible to everyone — without hype or fear.

This book is part of a broader effort to offer reliable, human-centered insights as the technology evolves.

So let's begin.

Where AI Stands Today

In just a few years, artificial intelligence has leaped from narrow chatbots to multimodal systems that can reason across text, images, and sound. Tools like ChatGPT-5 show how quickly the frontier is expanding — but the core ideas remain constant: data, pattern recognition, and human intention.

This book focuses on those timeless foundations so you can adapt to whatever comes next. For the latest examples, tools, and tutorials, visit masteringai.ai — updated regularly to keep your learning current.

Part I Understanding AI

Chapter 1

WHAT AI REALLY IS — AND WHY IT MATTERS RIGHT NOW

Introduction: Why This Really Matters Right Now

AI is no longer a distant concept — it's already woven into daily life. It's shaping the tools you use, the jobs you compete for, the decisions businesses make, and the opportunities available to your family. Whether you notice it or not, AI now influences the way you search, learn, shop, communicate, work, and create.

What's changing is the pace. For the first time, ordinary people — not just engineers — can tap into advanced AI tools to work smarter, learn faster, automate tasks, and unlock entirely new possibilities in their personal and professional lives. And the gap between those who understand AI and those who don't is growing quickly.

This book starts by grounding you in the essentials — what AI really is, what it isn't, and why it's transforming so much so quickly. Once the foundation is clear, we'll show you exactly how to use AI to improve your life, your work, your creativity, and your future.

---

1. How We Teach Machines to Think

Imagine a machine that doesn't just follow a fixed set of rules but actually learns from experience, adapts to new situations, and solves problems in ways that resemble human thinking. That's the promise — and the essence — of Artificial Intelligence.

Artificial Intelligence (AI) refers to the simulation of human intelligence in computer systems. It allows machines to perform tasks that typically require human thought, such as learning from experience, reasoning through problems, making decisions, and understanding language. These capabilities are driven by algorithms — structured sets of instructions — and powered by vast amounts of data. Together, algorithms and data enable machines to identify patterns, make predictions, and improve their performance over time.

---

2. From Simple Automation to Smart Systems

AI exists on a spectrum. On one end are simpler forms of automation — machines that follow predefined instructions to complete repetitive tasks reliably and efficiently. These systems don't learn or adapt; they just execute.

On the other end of the spectrum are more advanced models built using neural networks, inspired by the structure of the human brain, are the systems capable of deep learning — a type of AI that can recognize faces, understand spoken language, translate text, predict outcomes, and even generate original content — such as stories, artwork, and code.

This wide range of capabilities is what makes AI such a versatile and transformative force. From factory robots to language translators to creative tools, AI is reshaping how we interact with technology and with each other.

---

3. What Enabled the Rise of Modern AI?

While the idea of intelligent machines has existed for decades, several key technological shifts in recent years have accelerated AI's real-world impact:

- **The Internet** has given machines access to a massive, globally connected pool of information — text, images, data, and more — available at unprecedented speed and scale.

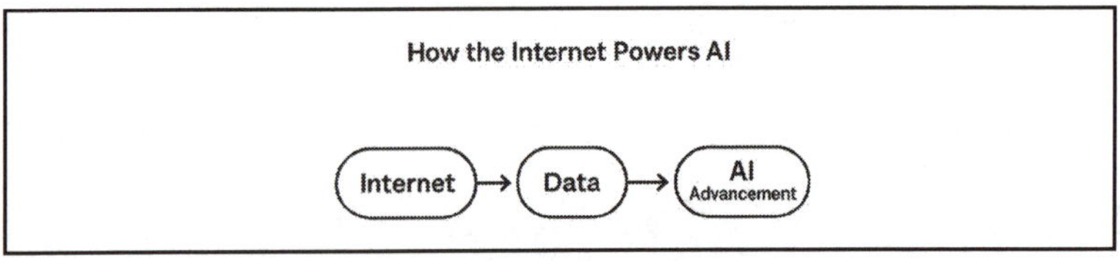

- Distributed computing allows data to be processed across many servers simultaneously, making it possible to analyze datasets that no single computer could handle alone.
- The Internet of Things (IoT) has linked everyday devices — from smartphones to smart refrigerators — into a network constantly generating real-time data.
- Social media platforms have exploded the availability of unstructured data — photos, videos, conversations, and behavior patterns — giving AI systems a rich source of raw input to learn from.

Together, these developments have created the perfect conditions for AI to move from theory to reality, enabling it to evolve rapidly and integrate into everything from healthcare and finance to entertainment, logistics, and personal devices.

---

4. AI vs. Human Intelligence

To understand AI more clearly, it helps to compare it with the original: the human mind.

Human intelligence is deeply intuitive. It involves critical thinking, emotional understanding, creativity, and the ability to adapt to completely new and unfamiliar situations. Humans combine logic and instinct, emotion and memory, to make decisions in complex and nuanced environments.

Artificial intelligence, by contrast, is designed to simulate certain aspects of that cognitive process. It can analyze enormous amounts of data far faster than a human ever could. It can learn patterns, follow logic, and make data-driven decisions. But it lacks emotional understanding, genuine creativity, and the kind of flexible intuition that humans use every day.

AI systems don't feel curiosity, empathy, or fear. They don't understand the meaning behind a joke or the emotion in a voice. They operate based on data and logic, not gut instinct or personal experience.

---

5. AI vs. Automation

AI is often lumped together with automation — and while they're related, they're not the same thing.

Automation is about speed and efficiency. It follows fixed rules and executes predefined tasks without deviation. Think of an assembly-line robot that bolts parts in place, or a script that sends out invoice emails every Friday. These systems don't "think" — they simply act according to a programmed routine.

AI, on the other hand, is about adaptation. It doesn't just follow instructions; it learns from data, adjusts its behavior, and makes decisions based on patterns. A chatbot that answers questions differently depending on how you phrase them — that's AI. A chatbot that gives the same response no matter what — that's automation.

Here's a simple way to remember the difference:

- **Automation** is rule-based.
- **AI** is data-driven and dynamic.

In practice, the two often work together. AI can enhance automation by making it smarter — adapting to new situations, personalizing experiences, or catching errors before they happen. But not every automated system involves AI, and not all AI solutions are designed for automation.

Understanding this distinction helps clarify what AI is really adding — not just doing things faster, but doing them more intelligently.

---

6. The Bottom Line: What AI Really Is

Artificial Intelligence isn't a mind, a personality, or a self-aware entity. It's a collection of techniques and tools that simulate aspects of human reasoning using data and computation. And while it's not "thinking" in the human sense, it's undeniably powerful — already transforming industries, improving decisions, and reshaping the digital world around us.

Understanding AI starts with seeing it clearly — not as science fiction or hype, but as a real-world tool with real strengths and clear limits. That's the foundation for the rest of this book.

Quick Takeaways: What Is AI?

» **AI simulates aspects of human intelligence.**

It enables machines to perform tasks like learning, reasoning, and decision-making—using algorithms and data instead of intuition or emotion.

» **AI exists on a spectrum.**

From simple automation to advanced deep learning systems, AI ranges from rule-based processes to neural networks that mimic how humans learn.

» **Modern AI is powered by data and connectivity.**

The rise of the internet, cloud computing, IoT, and social media has made vast, diverse datasets available for machines to learn from at scale.

» **AI is not human intelligence.**

AI can process information quickly and follow logic, but it lacks emotional depth, self-awareness, and creative intuition.

» **AI is a tool, not a mind.**

It doesn't think or feel — it identifies patterns and solves problems within defined limits, making it incredibly useful but not autonomous.

Chapter Summary

Artificial Intelligence is not science fiction — it's a real, evolving technology built to simulate parts of how we think. While it doesn't replicate the full richness of human intelligence, it's already reshaping how we work, interact, and innovate. Understanding what AI is — and what it isn't — is the first step toward using it wisely and navigating the future it's helping to build.

Chapter 2

DEBUNKING MYTHS AND MISCONCEPTIONS

Introduction: Debunking the Myths of AI

When most people hear the term artificial intelligence, their imaginations are instantly sparked by visions of sentient robots, evil supercomputers, or futuristic societies where machines have taken over the world. Hollywood has played a major role in shaping this image, filling screens with dramatic portrayals of AI — often thrilling, but rarely accurate. These stories can be entertaining, even thought-provoking, but they blur the line between science fiction and science fact.

To understand what AI really is — and what it isn't — we need to peel back the hype and confront the most persistent myths that cloud public perception.

By separating fantasy from reality, we can approach AI with clarity instead of fear or overconfidence.

---

1. The Hollywood Hype: Sentient Robots and Evil Supercomputers

From *The Terminator* to *Ex Machina,* AI is often depicted as a conscious, calculating force bent on humanity's destruction. These portrayals make for great cinema, but they have very little to do with the reality of AI today.

The systems we currently use—from voice assistants to fraud-detection tools — are examples of narrow AI, meaning they're designed to do specific tasks. They don't possess emotions, intentions, self-awareness, or desires. They're not plotting, deciding, or dreaming.

Even the smartest AI systems today can't easily shift between tasks the way humans can. An algorithm trained to identify tumors in X-rays can't suddenly learn to drive a car or write a novel. Its "intelligence" is limited to the task it was trained for — nothing more. This limitation is one of the biggest differences between real AI and the science-fiction versions people imagine.

2. The "AI is Magic" Fallacy: It's Not a Mysterious Black Box

Another myth paints AI as something mystical — an inscrutable black box that somehow "just works." This perception makes AI seem distant or unknowable, which can lead to fear, blind trust, or both.

But the truth is far less magical and far more logical.

At its core, AI is built on algorithms — step-by-step instructions — combined with large amounts of data. These systems learn patterns by analyzing examples and adjusting their responses based on feedback. It's similar to how a child learns: if you show a child enough pictures of cats and dogs, they'll eventually figure out the difference. AI learns in much the same way, only with far more data and without any understanding of what a "cat" or "dog" truly means.

Understanding this helps demystify AI and makes it more approachable. These are tools built by people, trained on data created by people, and following rules designed by people.

---

3. "AI Will Steal All Our Jobs" Oversimplification

One of the most common fears about AI is that it will wipe out entire industries and leave millions unemployed. It's true that automation will eliminate some roles, particularly those based on repetitive or routine tasks. But that's only part of the story.

Throughout history, every major technological shift has disrupted some jobs while creating new ones. The automobile displaced horse-drawn carriages but launched entire industries in manufacturing, logistics, infrastructure, and repair. The same pattern is unfolding with AI.

As some jobs become obsolete, new ones are already emerging in areas like:

- machine learning
- robotics maintenance
- AI ethics
- data annotation
- prompt engineering
- human-AI collaboration roles
- and industries we haven't even imagined yet.

Moreover, as machines take on routine tasks, human skills like creativity, empathy, communication, judgment, and strategic thinking become even more valuable.

In many cases, AI won't replace workers — it will **augment** them.

---

4. Other Common Myths Worth Debunking

"AI is Infallible"

Because AI is based on data and math, it's easy to assume it's always correct. But AI can make serious mistakes — especially when trained on flawed or biased data.

If an AI system learns from biased hiring records or incomplete medical histories, it may inherit and amplify those flaws.

This becomes especially dangerous in high-stakes areas like:

- healthcare
- criminal justice
- lending
- hiring

AI isn't inherently fair or objective. It reflects the data and assumptions it's built on — and both come from humans.

"AI Will Eventually Replace Humans Completely"

There's a dystopian narrative that envisions AI eventually making humans obsolete, running everything faster, cheaper, and better. But the most productive uses of AI today — and likely tomorrow — come from collaboration, not competition.

This idea is sometimes called augmented intelligence: machines that enhance human capabilities rather than replace them.

A doctor with an AI assistant that highlights potential abnormalities in X-rays can make faster, more accurate diagnoses.

A writer using AI for first drafts still applies their own voice and judgment.

Together, humans and machines outperform either one alone.

"Only Tech Experts Need to Understand AI"

Many people assume that AI is a technical domain best left to coders and engineers. But as AI touches more parts of everyday life — healthcare, finance, education, hiring, entertainment, transportation — understanding how AI works becomes a form of modern literacy.

You don't need programming skills to ask smart questions about:

- how AI systems make decisions
- what data they use
- whether they're fair
- how they affect your privacy and opportunities

The more diverse voices involved in shaping AI, the more human-centered these systems become.

---

5. The Bottom Line: Cutting Through the Hype

By separating fact from fiction, we can have more grounded, informed conversations about artificial intelligence. Understanding what AI actually is — and what it isn't — puts the power back in our hands. It allows us to engage with these technologies thoughtfully, use them wisely, and demand better systems that reflect our values.

AI isn't here to replace us or rule us. It's here to work alongside us, if we choose to use it that way.

The future of AI isn't written by machines — it's shaped by people, and that means all of us.

Quick Takeaways: Debunking AI Myths

» Today's AI doesn't think or feel.

Current AI systems process data to perform specific tasks. They aren't conscious, emotional, or self-motivated — unlike what Hollywood often portrays.

» AI isn't magic.

AI follows mathematical rules and statistical patterns. Understanding its foundations makes it less mysterious and more approachable.

» AI will reshape jobs, not erase them.

While automation will replace some roles, it will also create new industries and demand human skills like creativity, empathy, and complex problem-solving.

» AI can make mistakes.

AI is only as accurate as the data it's trained on. Biases in data can lead to flawed outcomes, making human oversight essential.

» **The future is human-AI collaboration.**

AI enhances human abilities rather than replacing them. Augmented intelligence — humans and machines working together—is the real opportunity.

» **AI literacy matters for everyone.**

You don't have to be a programmer to understand AI. Knowing how it affects decisions, rights, and opportunities is part of being an informed citizen today.

Chapter Summary

AI is surrounded by hype, misconceptions, and cultural myths — but the truth is both simpler and more meaningful. It's not magic, it's not sentient, and it's not here to replace us. It's a powerful tool built by people, shaped by data, and guided by human choices. When we understand what AI really is — and what it's not — we can approach it with clarity, use it responsibly, and ensure it benefits everyone, not just a few.

When we separate myth from reality, we gain the confidence to use AI wisely rather than fear it.

Chapter 3

WIRED FOR CHANGE: THE REAL IMPACT OF AI

Introduction: Why AI Matters Now

Artificial Intelligence is no longer science fiction. It's not tucked away in research labs or hidden behind complex algorithms only a few experts understand. AI is here, and it's reshaping how we live, work, and think — whether we're ready or not.

But why does it matter so much?

Because AI isn't a future technology. It's a present-day force — one that influences your job, your choices, your information, and your opportunities right now.

---

1. AI Changes How We Work

AI isn't just about robots on factory floors. It's about the software that reads X-rays faster than radiologists, the chatbots that handle customer service, and the algorithms that screen job applications. Tools like Grammarly, ChatGPT, and automated research assistants help professionals write, analyze, and create faster and more effectively. Logistics companies like UPS use AI to optimize delivery routes, saving millions of dollars annually.

Some jobs are being redefined. Lawyers use AI to sift through case law. Financial analysts use it to predict market trends. Journalists use it to generate data-driven stories.

» **The workers who adapt to AI will thrive. Those who don't risk falling behind — not because AI "replaces" them, but because others will use AI to work faster, think bigger, and produce more.**

And it's not just specialized fields. From marketing to healthcare, AI tools are becoming everyday companions in almost every profession.

---

2. AI Impacts Daily Life

AI doesn't just change how we work — it's also deeply embedded in our everyday routines, often in ways we barely notice. When Netflix recommends a new series you might like, or Google Maps adjusts your route because of a traffic jam ahead, or your Instagram feed seems oddly in sync with recent conversations — that's AI analyzing patterns and making decisions in real time. It powers the digital experiences we rely on every day.

But AI doesn't just offer convenience. It also subtly shapes what we see, how we shop, and even what we believe. Algorithms decide which news articles show up in your feed, which ads you're shown, and which videos autoplay next. Over time, these small nudges can have a large influence — not just on our choices, but on our opinions, values, and behavior.

Understanding these digital patterns helps you stay in control of your online environment — instead of being steered unconsciously by systems operating in the background.

---

3. AI Raises Big Ethical Questions

With so much power, AI inevitably raises serious ethical questions. These systems are only as good — or as dangerous—as the data they're trained on. When facial recognition software misidentifies people of color at disproportionately high rates, the consequences go far beyond bad predictions; they become issues of justice, equity, and civil rights. And when AI-generated deepfakes spread false information online, the very idea of truth becomes harder to protect.

Governments and companies alike are scrambling to catch up. In 2023, the European Union passed the AI Act to regulate high-risk AI systems. Companies like OpenAI and Google DeepMind have called for greater transparency, oversight, and international cooperation on AI ethics and safety. But the technology is moving fast — often faster than our policies and institutions.

Ethical AI won't build itself. It requires informed citizens, public dialogue, strong standards, and collective pressure to ensure AI serves the many, not just the few.

---

4. AI Shapes the Future

Beyond its current impact, AI is already helping define what the future looks like. Scientists are using AI to design new proteins that could one day lead to cures for diseases like Alzheimer's. Climate researchers are building AI models that can predict extreme weather events more accurately, helping governments and communities prepare for disasters and adapt to change. In transportation, autonomous vehicles could reduce traffic fatalities by removing human error — a transformation that could save thousands of lives.

But with great potential also comes real risk. Military drones with autonomous targeting capabilities raise complex ethical and legal questions. Mass surveillance powered by AI threatens privacy and civil liberties, especially in countries without strong protections in place.

Whether AI becomes a force for global good or a tool for control depends not just on the technology itself — but on the guardrails we put in place and the choices society makes now.

---

5. AI is for Everyone

For a long time, AI seemed like it belonged exclusively to tech giants, academic labs, and elite developers. That's no longer true. Today, the tools are becoming democratized. Anyone can explore AI with little more than a browser. Platforms like Teachable Machine let you build simple models without a single line of code. Creative tools let you make music, generate art, write stories, or even launch a business — with AI acting as a co-pilot rather than a replacement.

Understanding AI doesn't mean becoming a programmer. It means becoming literate in a world where AI influences your job prospects, your healthcare decisions, your news consumption, and your civic rights.

And the more diverse the voices involved in shaping AI — artists, educators, parents, frontline workers — the more likely we are to build systems that are fair, useful, and truly reflective of human needs.

---

6. The Bottom Line: Why AI Matters

AI is already touching nearly every part of our lives — from the work we do to the choices we make, the information we trust, and the future we're building. It's no longer happening quietly in the background; it's right here, reshaping the present and influencing what comes next.

Ignoring AI isn't just risky. It means leaving decisions about your future in someone else's hands. Engaging with AI doesn't require a technical background — only curiosity, awareness, and the willingness to understand how these systems work and what they mean for you.

You don't need to write code to be part of the conversation.

You just need to participate.

Quick Takeaways: The Real Impact of AI

» **AI is transforming work.**

It automates tasks, enhances productivity, and redefines roles across industries — from logistics to journalism to law.

» **AI is embedded in daily life.**

It powers the recommendations, navigation, ads, and content we see — often without us noticing.

» **AI raises ethical stakes.**

From algorithmic bias to deepfakes to surveillance, its misuse can impact civil rights, truth, and public trust.

» **AI is shaping the future.**

It's accelerating breakthroughs in science, healthcare, climate modeling, and transportation — while also raising serious risks around warfare and privacy.

» **AI is for everyone.**

You don't need to code to engage with AI. Understanding how it works gives you agency in a world increasingly influenced by automated systems.

Chapter Summary

AI is no longer hidden in labs or confined to futuristic visions — it's already transforming how we work, live, and connect. From automating tasks to influencing daily choices, from raising ethical challenges to creating new opportunities, AI is becoming a defining force of our time.

Understanding its impact isn't optional. It's essential for navigating a world where AI shapes not just industries, but societies — and individual lives. Staying curious, informed, and engaged is how we ensure AI evolves in ways that benefit everyone.

Chapter 4

A BRIEF HISTORY OF AI: HOW WE GOT HERE

Introduction: From Ancient Tools to Thinking Machines

The story of artificial intelligence is tightly woven into the broader history of human innovation. Since ancient times, people have tried to create tools that could ease not just physical labor but also mental effort. The abacus, invented over 2,000 years ago, and early mechanical calculators were some of humanity's first steps toward automating thinking.

But the formal idea of AI — machines that could "think" — didn't truly take off until the mid-20th century. That's when computing and imagination collided to give birth to a revolutionary concept.

And what followed was a journey of bold dreams, setbacks, breakthroughs, and reinvention — a cycle that continues today.

---

1. Early Beginnings: The Dream of Thinking Machines

The 1950s mark the official starting line for AI as a scientific field.

British mathematician Alan Turing posed a powerful question: Can machines think? His 1950 paper, "Computing Machinery and Intelligence," introduced what we now call the Turing Test — a way to measure whether a machine can exhibit behavior indistinguishable from a human.

Just a few years later, John McCarthy — often called the "father of AI" — coined the term artificial intelligence and organized the Dartmouth Workshop in 1956. This workshop brought together early pioneers like Marvin Minsky and Claude Shannon. Their optimism was bold — they believed that building a truly intelligent machine might only take a few decades.

The 1960s produced early AI programs like ELIZA, which mimicked a psychotherapist using simple pattern matching, and SHRDLU, which could manipulate virtual blocks through typed commands. These were primitive, but they showed that machines could begin to interact using human language.

However, progress wasn't as smooth as hoped. AI soon ran into tough problems: limited computing power, lack of data, and the realization that human reasoning is far more complex than initially thought.

This led to periods of deep disillusionment known as AI Winters — funding dried up, excitement faded, and research slowed to a crawl.

But the core idea never disappeared. Scientists simply needed a new approach.

---

2. The Rise of Machine Learning: Shifting from Rules to Data

In the 1980s and 1990s, a major shift began. Instead of trying to hard-code intelligence with a fixed set of rules, researchers moved toward machine learning — giving machines the ability to learn patterns from data without being explicitly programmed for every scenario.

Expert systems, like XCON at Digital Equipment Corporation, showed promise by using rule-based logic to make decisions, especially in fields like medical diagnosis or industrial planning. But expert systems had limits: they were brittle, hard to maintain, and struggled with real-world complexity.

Then came neural networks, inspired by the structure of the human brain. Early versions were limited, but they introduced a crucial idea: intelligence could emerge through training, not just programming.

As computing power grew and more digital data became available, machine learning began to rise — setting the stage for the next revolution.

AI wasn't just being programmed.

It was beginning to learn.

---

3. The Deep Learning Revolution: Learning by Layers

The 2000s and 2010s brought a massive leap forward, powered by deep learning — a type of machine learning that uses multi-layered neural networks to solve complex tasks.

A breakthrough came in 2012 with AlexNet, a deep neural network trained on the ImageNet dataset. It shattered previous benchmarks for image recognition and kicked off the modern AI boom.

It wasn't just about bigger datasets or faster GPUs. Techniques like backpropagation allowed these networks to refine themselves layer by layer, improving through feedback.

Natural language processing made similar leaps. Google Translate shifted from rule-based systems to neural machine translation, producing more natural and fluent translations.

Then came generative models like GPT, which could produce human-like text from a simple prompt — transforming how we think about language, creativity, and digital interaction.

The era of deep learning turned AI from a specialized tool into a mainstream technology.

---

4. Today and Tomorrow: AI Is Everywhere

Today, AI is embedded in the tools and systems we use every day.

- Voice assistants like **Alexa** and **Siri**
- Recommendation engines on **Netflix**, **YouTube**, and **Spotify**
- Fraud detection in banking
- Diagnostic tools in healthcare
- AI-powered chatbots and content generators

And the future is already taking shape:

- **Multimodal AI** that can process text, images, and audio together
- **Autonomous systems** like self-driving cars and intelligent drones
- **Ethical AI** and **AI governance** rising in importance as we confront bias, misinformation, and surveillance risks
- **Hybrid models** that combine neural networks with reasoning, memory, and explainability

We're also seeing systems that can not only perform tasks but also justify or explain their reasoning — a major step toward AI people can trust.

In other words: AI is no longer a research topic.

It's infrastructure.

---

5. Wrapping Up: Why the Past Powers the Future

The history of AI is a story of bold dreams, unexpected setbacks, and stunning breakthroughs. From the early visions of Alan Turing to today's generative and multimodal systems, AI's journey has been anything but linear.

Understanding where AI came from helps us appreciate not just its potential, but also its pitfalls. It reminds us that AI is not magic. It's the result of decades of human effort — and it's still evolving.

In the next chapters, we'll explore how AI works today, what it can and can't do, and how it's shaping the world we're living in right now.

---

6. Quick Timeline: Key Moments in AI History

▸ **Ancient Times**

- Invention of the abacus and early mechanical calculators — first steps toward automating thinking.

▸ **1950**

- **Alan Turing** proposes the Turing Test to measure machine intelligence.

▸ **1956**

- **Dartmouth Workshop**: John McCarthy coins the term *artificial intelligence*.
- Birth of AI as a scientific field.

▸ **1960s**

- Early programs like **ELIZA** (simulated conversation) and **SHRDLU** (manipulated objects in a virtual world).

▸ **1970s–1980s**

- **AI Winters**: Funding and interest decline due to unmet expectations.

▸ **1980s**

- Rise of **Expert Systems** (e.g., XCON), using rules to mimic human decision-making.

▸ **1990s**

- Neural networks return.
- Growth of **machine learning**: shifting focus from rules to learning patterns from data.

▸ **2012**

- **AlexNet** wins ImageNet competition, crushing previous records and igniting the **deep learning revolution**.

▸ **2010s**

- Explosion of AI applications in NLP (e.g., Google Translate) and computer vision.
- Launch of models like **GPT** by OpenAI.

▸ **2020s**

- AI moves into real-world applications: self-driving cars, medical AI, chatbots.
- Growing focus on **Ethical AI** and **AI governance**.

▸ **Present and Beyond**

- Multimodal AI, hybrid models, AI explainability, and tighter integration into daily life.

Quick Takeaways: A Brief History of AI

» The Idea Isn't New

The dream of intelligent machines dates back decades—even centuries—but AI as a scientific field began in the 1950s with pioneers like Alan Turing and John McCarthy.

» From Rules to Learning

Early AI relied on hard-coded logic and struggled with real-world complexity. The breakthrough came when researchers shifted to machine learning—letting AI learn from data instead of explicit rules.

» Progress Wasn't Linear

AI faced major setbacks during the "AI Winters" when expectations outpaced reality. These periods remind us that hype alone can't sustain progress—real breakthroughs take time.

» Deep Learning Changed Everything

Starting in 2012, deep neural networks began outperforming traditional methods in image recognition, language translation, and more—igniting the modern AI boom.

» AI Is Now Part of Everyday Life

From Netflix recommendations to self-driving cars, AI is no longer an experiment—it's embedded in how we work, play, and connect.

» We're Still Writing the Story

As AI evolves, new challenges emerge: ethical concerns, hallucinations, privacy, and control. Understanding the past helps us navigate the road ahead wisely.

Chapter Summary

AI's history is not a straight line — it's a loop of ambition, setbacks, and breakthroughs. From Turing's earliest questions to today's generative models, the field has transformed how we think about intelligence, creativity, and possibility.

Understanding where AI came from gives us the context to navigate where it's going — and why it matters more than ever.

Chapter 5

THE EVOLUTION OF AI: TRADITIONAL AI VS GENERATIVE AI

Introduction: A Technological Turning Point

Generative AI is everywhere — in headlines, in boardrooms, and increasingly in the tools we use every day. It can write, draw, compose, and even simulate conversation with surprising fluency.

But a common question remains:

» **How is generative AI different from the AI we used 10, 20, or even 30 years ago?**

To understand that, we need to revisit where AI started — and what it was built to do. Traditional AI focused on narrowly defined tasks, relying on structured data, handcrafted rules, and slow feedback loops. It was powerful but limited.

Then came a new kind of intelligence: generative models trained on internet-scale datasets, capable of creating rather than just predicting. This shift didn't just make AI faster or bigger — it changed its architecture, its purpose, and its role in society.

In this chapter, we'll explore the core differences between traditional and generative AI, comparing their data, design, capabilities, and implications. By the end, you'll see why today's AI tools feel fundamentally different — and what that means for the future of work, creativity, and decision-making.

---

1. Traditional AI: Predictive and Pattern-Based

The Old Architecture

Traditional AI systems were built on three core layers:

1. **Repository**

   A database where all internal information was stored: tables, images, documents, customer records.

2. **Analytics Platform**

   Tools like **Alteryx Designer** or **Azure Machine Learning Studio** were used to develop forecasting models.

   Example: Predicting which products would sell out during a holiday season.

3. **Application Layer**

   Once the sales forecasting model was ready, it was integrated into the inventory management software. The system could then automatically adjust stock orders based on predicted product demand.

Feedback Loop: Learning from Experience

To make it true AI (not just prediction), a **feedback loop** was added.

When a model was wrong — predicting someone would cancel when they didn't — the system learned, improved, and became smarter over time. The mantra was simple:

» **Fool me once, shame on you. Fool me twice, shame on me.**

Traditional AI could learn, but it was slow, narrow, and heavily dependent on internal company data. Each system was designed for a specific purpose — and changing that purpose often meant rebuilding models from the ground up.

---

2. Generative AI: A Whole New Paradigm

How the New Architecture Changes Everything

Generative AI isn't just an evolution of traditional AI. It represents a **complete shift** in how machines learn, reason, and create. To understand this shift, let's break down the key architectural differences that define this new wave.

- **Data Source**

Traditional AI systems were built primarily on internal, organization-specific data — structured tables, transaction records, customer histories, and operational logs.

Generative AI, in contrast, draws from enormous, open-ended datasets gathered from across the internet and other large-scale sources. This includes books, academic papers, code repositories, product manuals, blogs, and online conversations. This massive and diverse pool of knowledge allows generative models to engage with a wide range of topics and tasks.

However, it also introduces challenges: data noise, outdated information, and embedded biases — making careful prompting, critical thinking, and human oversight essential.

• Model Size

In traditional AI, models were purpose-built: small to medium in size, carefully tailored to solve specific problems like predicting customer churn or detecting fraud.

Generative AI models — such as GPT, Claude, or Gemini — are massive in scale, often containing billions or even trillions of parameters. Training these models requires weeks or months of computation on powerful clusters or supercomputers.

Their sheer size enables them to recognize complex patterns, generate human-like responses, and perform a wide variety of tasks — all from a single unified architecture.

• Customization

Traditional AI demanded a manual setup: feature engineering, rule writing, and model construction all performed by data scientists and domain experts.

Generative AI is customized through **prompting and fine-tuning**.

- **Prompting** involves carefully crafting inputs that guide the model to produce relevant, accurate outputs.
- **Fine-tuning** retrains the model on domain-specific data to align it even more closely with a specific industry, company, or task.

This shift means organizations can adapt large, general-purpose models to their needs **without rebuilding everything from scratch**.

• Learning Process

Generative AI introduces a more **fluid, interaction-driven learning process**. Although these models don't retrain themselves automatically after deployment, their behavior can be rapidly shaped through:

- better prompts
- reinforcement-style feedback
- iterative tuning cycles

This enables constant evolution in how the model is used — without a full technical rebuild each time.

• Application Focus

Traditional AI excelled at **narrow, well-defined tasks**: diagnosing diseases from scans, spotting fraudulent transactions, predicting machine failures — and typically, each model did one job.

Generative AI, however, is designed for **broad, creative functions**. It can summarize documents, draft reports, write code, design marketing copy, compose music, generate images, and simulate human-like conversation.

Its role isn't just prediction or classification — it's **creation, synthesis, and communication** at a scale traditional AI could never reach.

---

3. Why It Matters

Traditional AI was powerful in its time, but it was bounded by clear limitations: access to company-specific data, narrow model design, and labor-intensive tuning. It worked best behind the scenes — improving logistics, crunching numbers, spotting anomalies — but always within strict confines.

Generative AI breaks those walls.

It's not just an evolution — it's a paradigm shift. These models aren't limited to structured enterprise data or one job at a time. They're trained on global-scale information — encompassing human language, scientific knowledge, culture, code, and more — and designed to adapt across many domains.

This leap changes the nature of what AI can be:

- **From assistant to collaborator:** AI can now brainstorm with writers, debug with coders, and compose music with musicians.
- **From reactive to proactive:** Instead of waiting for structured inputs, generative AI can engage with open-ended prompts, suggest next steps, or even anticipate needs.
- **From niche to ubiquitous:** AI is moving from specialized applications into everyday life — showing up in classrooms, creative studios, marketing departments, hospitals, and homes.

But with this power come new challenges:

- **Ethical complexity:** More capable systems raise more difficult questions — about fairness, authorship, surveillance, and misuse.
- **Loss of explainability:** As models grow in complexity, it gets harder to trace their reasoning — making trust and transparency vital.

- **Information overload:** With the ability to generate endless content, the value now lies in what we ask, how we verify, and why we use AI at all.

Understanding this shift is no longer optional. Whether you're a student, entrepreneur, artist, educator, or policymaker — knowing the difference between traditional and generative AI helps you:

- Make smarter choices
- Anticipate risks
- Unlock new opportunities

This is the moment AI stopped being invisible infrastructure — and became a visible partner in human decision-making, creation, and knowledge itself.

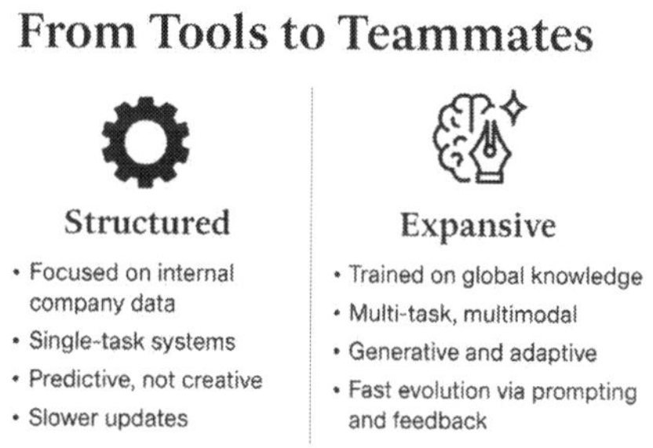

---

4. Comparing Traditional and Generative AI: A Side-by-Side View

You can think of the differences in five key dimensions:

Feature	Traditional AI	Generative AI
Data Source	Company-specific repositories	Massive global datasets
Model Size	Small to medium, task-specific	Enormous, general-purpose
Customization	Built from scratch	Prompting and fine-tuning
Learning Process	Manual tuning + feedback	Prompt optimization + feedback
Application Focus	Narrow, specific tasks	Broad, flexible generation and creation

In simple terms:

Traditional AI and Generative AI are not just different in scale — they're fundamentally different in spirit. Here's a final analogy to capture the difference:

- Traditional AI is like a chef carefully cooking one meal at a time based on past recipes.
- Generative AI is like having an entire library of every recipe ever written — and inventing new dishes on the fly.

Quick Takeaways: Traditional AI vs Generative AI

» Data Source

Traditional AI uses company-specific data; Generative AI pulls from the global internet.

» Model Size

Traditional models are small; Generative models are enormous and multi-purpose.

» Customization

Traditional AI requires manual setup; Generative AI adapts through prompting and fine-tuning.

» Learning Process

Traditional AI updates are manual; Generative AI evolves fluidly through prompt optimization.

» Application Focus

Traditional AI is task-specific; Generative AI is broad, creative, and multi-talented.

Chapter Summary

Generative AI represents more than just an upgrade to traditional AI — it marks a complete architectural and conceptual shift. Where traditional AI was narrow, static, and task-specific, generative AI is broad, dynamic, and creative.

It draws from global-scale datasets, operates at massive scale, and adapts through prompting instead of being rebuilt from scratch. This new model isn't just more powerful; it's fundamentally different in how it learns, functions, and impacts our world.

Understanding this evolution is key to grasping where AI is going — and what it means for everything from business to creativity, education to ethics.

Chapter 6

TYPES OF AI:
FROM SIMPLE SYSTEMS TO ADVANCED INTELLIGENCE

Introduction:
AI Is Not One Thing — It's Many Different Models

Artificial Intelligence is often discussed as if it were a single technology — one unified idea with one set of capabilities. But the truth is far more interesting. AI comes in many forms, built in different ways, serving different purposes, and behaving at different levels of "intelligence."

- Some systems follow fixed rules.
- Some learn from data.
- Some predict.
- Some create.
- Some reason only about the present moment.
- Others build models of the past to improve over time.

Understanding AI requires understanding these distinctions. This chapter brings clarity by breaking the landscape into a simple, practical framework used across the AI community — one that separates **what AI can do, how it works**, and **how broad its intelligence really is**.

Once you understand these layers, the rest of the book — and the broader world of AI — will make far more sense.

PART 1 — The Three Ways Experts Classify AI

These are the three lenses used by researchers, industry, and policymakers to describe AI. Each one answers a different question:

- **Capability:** How intelligent or autonomous is the system?
- **Function:** What is the AI designed to do?
- **Technique:** How does the AI work internally?

These categories overlap in real systems — a single AI model can belong to one capability level, one functional type, and one technical method.

Together, these three lenses give us a complete picture.

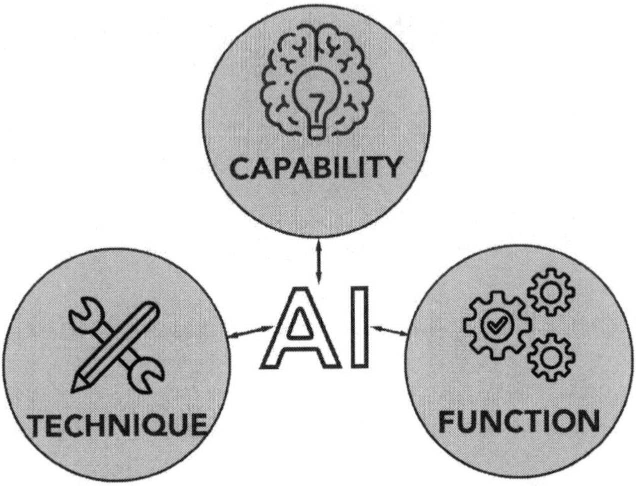

---

1. Classifying AI by Capability: How "Smart" the System Is

This classification describes AI based on its cognitive abilities — how it processes information, adapts, or (theoretically) understands.

There are four categories, moving from the simplest systems to the most conceptual.

Reactive AI

Reactive AI is the simplest and oldest form. These systems respond to inputs with pre-programmed rules. They have:

- no memory
- no ability to learn
- no sense of context

They analyze the present moment and react instantly — nothing more.

Examples: classic chess engines, rule-based chat scripts

Core traits: predictable, fast, but inflexible

Limited Memory AI

This is the dominant form of AI in use today. These systems learn from data, store past information, and improve over time.

Self-driving cars, recommendation engines, language models, fraud detectors — all fall into this category.

Examples: ChatGPT, Netflix recommendations, self-driving car vision systems

Core traits: learns patterns, adapts, improves with data

This is where most modern breakthroughs occur.

Theory of Mind AI (Research Stage)

This is a forward-looking concept describing AI that could understand human emotions, intentions, and mental states.

It does not exist yet — but researchers are exploring early foundations.

Theory of Mind AI aims for:

- basic social awareness
- understanding that humans have beliefs and emotions
- adjusting behavior based on user context

Examples: experimental emotional AI tools

Core traits: conceptual empathy, anticipatory responses (in theory)

But current systems do not truly "understand" human emotion — they only detect patterns.

Self-Aware AI (Purely Theoretical)

This represents the philosophical edge of AI — systems with consciousness, subjective experience, or genuine self-awareness.

Researchers discuss it conceptually through ideas such as:

- metacognition — thinking about one's own thinking
- internal self-models — understanding oneself as an agent
- intentionality — forming independent goals
- persistent identity — self-awareness over time

No such system exists, and it may require breakthroughs beyond current AI science.

Examples: none

Core traits: hypothetical consciousness, theoretical autonomy
This category exists to frame ethical and philosophical discussions — not real technologies.

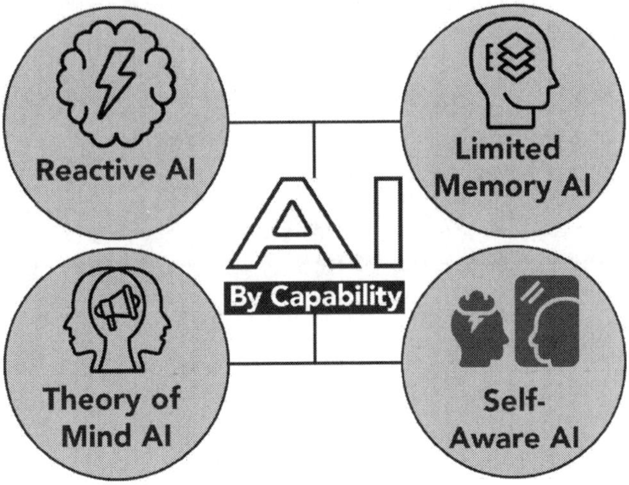

---

2. Classifying AI by Function: What the AI Is Designed to Do

This classification is widely used in business, analytics, and applied AI.

It groups AI systems by the type of output they produce — analyzing, predicting, recommending, or creating.

Descriptive (Diagnostic) AI

It helps us understand the past and make sense of data we already have.

Used for:

- spotting patterns
- identifying trends
- finding root causes
- summarizing historical performance

Examples:

- dashboards and analytics platforms
- customer-service complaint analysis
- quality-control monitoring

Key abilities:

- recognizes patterns in large datasets
- groups and labels data
- surfaces anomalies
- identifies what happened and why

Descriptive AI is the foundation of all analytics. It doesn't predict or recommend — it explains.

Predictive AI

Uses patterns in historical and real-time data to estimate what is likely to happen next.

Used for:

- sales forecasting
- churn prediction
- fraud detection
- medical risk scoring
- demand planning

Examples:

- predicting which customers might cancel
- estimating which machines will fail soon
- forecasting inventory spikes

Key abilities:

- forecasting future events
- clustering and classifying users
- probability scoring
- identifying early warning signals

Predictive AI is powerful — but it still stops short of telling us what to do with the predictions.

Prescriptive AI

Answers the question: "Given the situation, what should we do?"

Used for:

- process optimization
- personalization

- scheduling and logistics
- decision support

Examples:

- suggesting optimal delivery routes
- recommending the next-best product
- adjusting energy usage in smart buildings

Key abilities:

- optimization
- scenario simulation
- risk mitigation
- decision recommendations
- turns predictions into recommended actions

Prescriptive AI blends prediction and strategy — turning insight into action.

Generative (Cognitive) AI

Creates new content — not just numbers or forecasts, but words, images, audio, code, and ideas.

This is the newest — and most transformative — category of AI.

Used for:

- writing (emails, articles, scripts)
- creative work (art, design, music, video)
- coding and software development
- research summarization
- simulation and ideation
- customer support and conversation

Examples:

- ChatGPT drafting documents or explaining concepts
- Midjourney generating design concepts
- GitHub Copilot writing code
- Runway generating videos

Key abilities:

- content creation across formats (text, images, audio, video, code)
- summarization and explanation

- ideation and brainstorming
- style adaptation (write like a lawyer, designer, marketer)
- multi-step reasoning
- generating synthetic data for training

Why it matters:

Generative AI doesn't just analyze or recommend — it produces. But **generative AI** can still make errors — it's confident, not always correct — which is why human oversight matters.

It acts like a collaborator that can:

- propose ideas
- draft first versions
- refine your work
- speed up learning and exploration

It is the first type of AI that feels interactive, conversational, and creative — which is why it's changing industries so fast.

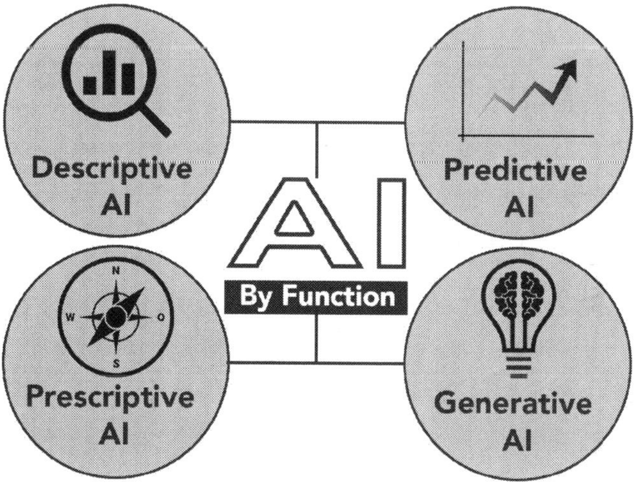

---

3. Classifying AI by Technique: How the AI Works Internally

This lens explains how AI makes decisions — the underlying method it uses to process information and learn.

Rule-Based AI — The "If–Then" Engine

The earliest and simplest form of AI.

How it works:

Humans write explicit rules.

If X happens → do Y.

No learning, no adaptation.

Analogy: A vending machine. Insert money → product comes out.

Used for: automation, compliance, decision trees

Example: A system that flags any expense above $1,000.

Strengths: predictable, transparent

Limitations: breaks easily when conditions change; cannot learn

Machine Learning (ML) — Learning Patterns from Data

Instead of rules, ML learns from examples.

How it works:

You give the system many labeled examples (e.g., "spam" / "not spam").

The system learns patterns and creates its own rules.

Analogy: A student learning by practicing hundreds of examples.

Used for: fraud detection, predictions, recommendations

Example: Spotify recommending music based on your listening habits.

Strengths: adaptable, improves with data

Limitations: performance depends heavily on training quality; brittle with unfamiliar data

Deep Learning (DL) — Many Layers, High Complexity

A powerful subset of ML using large neural networks.

How it works:

Instead of a single model learning simple patterns, DL uses dozens to hundreds of layers — each detecting increasingly abstract features (edges → shapes → objects → meaning).

Analogy: A giant team of specialists, each layer adding its own insight.

Used for: vision, speech, language, autonomous driving

Example: Analyzing X-rays to detect signs of cancer.

Strengths: excellent accuracy on complex, high-dimensional data

Limitations: harder to interpret ("black box"), requires massive data and compute

Generative AI — Building New Content from Learned Patterns

A specialized branch of deep learning focused on creation rather than classification.

How it works:

The model learns the structure of language, images, or audio — then generates new content that follows those patterns.

Analogy: A chef who has tasted every recipe in the world creating new dishes inspired by them.

Used for: writing, design, coding, simulation

Examples: GPT, Claude, Midjourney, Runway

Strengths: highly creative, flexible, multi-purpose

Limitations: may generate incorrect or biased content; requires human oversight

Here's the simplest way to think about it:

Rule-Based AI uses rules.
Machine Learning uses data.
Deep Learning uses layers.
Generative AI uses patterns to create.

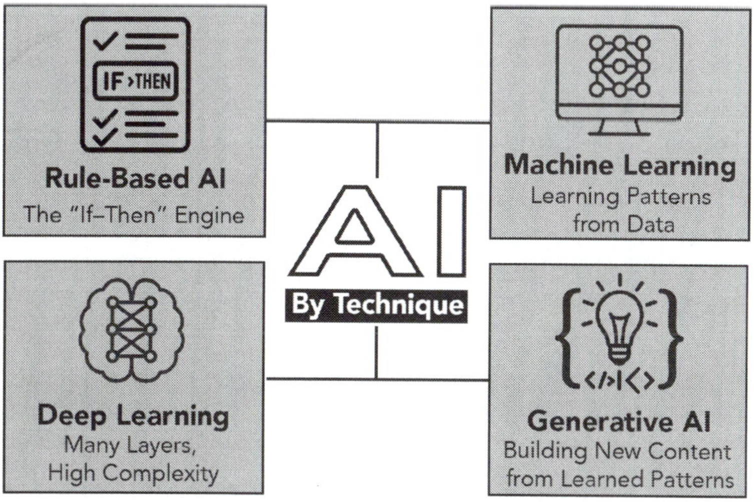

PART 2 —

The AI Scope Spectrum: What Exists vs. What's Coming

Beyond capabilities, functions, and techniques, there's another important question:

→ **How broad is the AI's intelligence?**

This is what researchers call the "AI spectrum," ranging from today's narrow, task-focused systems to hypothetical forms of machine intelligence that could one day match — or exceed — human cognition.

Narrow AI (Weak AI)

The AI we use today.

Powerful, practical, but specialized.

Examples: translation tools, facial recognition, recommendation engines, chatbots

Traits:

- Highly accurate within one domain
- Cannot generalize beyond its training
- No self-awareness or broad reasoning
- Even advanced LLMs like ChatGPT fall into this category

Narrow AI is excellent at one thing, but cannot apply that skill to unrelated tasks.

General AI (Strong AI)

A hypothetical system with human-level flexibility and reasoning.

It could learn any task, switch between them, and adapt in real time.

Examples: none yet

Traits:

- Learns across domains
- Understands context broadly
- Performs new tasks without retraining

This remains a research goal—not a reality.

Super AI

A speculative form of intelligence that surpasses human abilities in every cognitive domain — problem-solving, creativity, emotional understanding, and strategic reasoning.

Examples: none

Traits:

- Beyond human capability
- Entirely theoretical
- Central to ethical, philosophical, and safety debates

Super AI shapes long-term discussions, but has no prototypes today.

Why This Spectrum Matters

Understanding the AI spectrum helps separate today's reality from future possibilities.

Most concerns about AI "taking over" come from confusing narrow AI with the still-theoretical ideas of General or Super AI.

Knowing the difference grounds our expectations — and our decisions.

PART 3 — Why These Distinctions Matter

Understanding these classification systems helps clear up the confusion surrounding AI.

- People often mix up:
- capability (how smart it is)
- function (what it does)
- technique (how it works)
- scope (how broad its intelligence is)

A generative AI can be:

- narrow (scope),
- deep learning (technique),
- generative (function),
- limited memory (capability).

These categories overlap — and that's normal.
But when people misunderstand them, fear and hype grow.

Clear understanding helps you:

- evaluate tools realistically
- spot misinformation
- understand limitations
- anticipate risks
- make better personal and professional decisions

Quick Takeaways: Types of AI

» AI can be classified three ways

By capability, function, and technique. These three views help us understand what an AI system can do and how it does it.

» Most real-world systems are Limited Memory AI

They learn from data and improve over time, but they lack emotion, understanding, or self-awareness.

» Function matters

AI systems can be descriptive, predictive, prescriptive, or generative — and each serves a different role in solving problems.

» Technique shapes capability

Rule-based systems, machine learning, deep learning, and generative models each offer different strengths and limitations.

» Narrow AI dominates today's world

Even advanced tools like ChatGPT operate within specific boundaries — they are powerful but not "general" intelligence.

» General and Super AI remain theoretical

These concepts often appear in media or futuristic discussions but do not exist today and should not be confused with current AI technologies.

Chapter Summary

AI is not a single technology — it's a layered landscape with systems that vary in purpose, intelligence, and design. This chapter breaks AI into three essential dimensions: capability, function, and technique. By understanding these distinctions — and the broader spectrum from narrow AI to hypothetical strong and super AI — we gain a clearer, more grounded view of what AI can and cannot do today. That clarity helps us navigate a world where AI is everywhere, shaping decisions, behaviors, and opportunities. Understanding the landscape is the first step toward using AI wisely, responsibly, and confidently.

Chapter 7

HOW AI THINKS
ALGORITHMS, DATA, AND LEARNING

Introduction: How Machines "Think" Without Thinking

When people hear "artificial intelligence," they often picture movie scenes: machines holding conversations, solving complex problems, or acting with human-like emotions. It's easy to assume AI thinks the way we do. It doesn't.

AI doesn't imagine, reflect, or feel. It has no self-awareness or intentions. What it does have is the ability to process huge amounts of data, follow logical rules, and detect patterns at a speed and scale far beyond human capability.

At its core, an AI system is built to spot relationships in data and use those relationships to make decisions, classify information, or generate content. It doesn't truly "understand" the world — but it can model parts of it through algorithms, training processes, and feedback loops. In other words, AI learns from experience, but not in the way people do. It doesn't learn from intuition or emotion; it learns from examples, statistics, and rules.

To really grasp how AI works, we need to strip away the myths and get familiar with the mechanics. How do algorithms operate? What kinds of learning exist? Why is data so important — and what happens when that data is flawed?

This chapter walks through the foundational ideas behind how AI "thinks" — starting with data, moving into learning methods, and ending with feedback, human oversight, and the limits of what machines can do. You won't need math or programming to follow along — just a willingness to break big ideas into simple parts. Once you see how machines learn, the mystery of AI becomes much more manageable.

---

1. Data Is the Fuel of AI

If there's one thing AI absolutely depends on, it's data. Just as humans learn by observing the world and drawing lessons from experience, AI learns by processing examples — sometimes millions of them. But while people can learn from a few experiences and make intuitive leaps, AI systems usually need far more repetition. They don't guess the way we do. They improve only when they're trained on a large, well-organized supply of information.

You can think of data as the raw experience AI uses to "learn." If you want an AI to recognize cats, you don't show it just one photo. You feed it thousands — cats in different poses, lighting, colors, backgrounds, and angles. Over time, the AI detects shared patterns: the shape of the ears, the distance between the eyes, the curve of the tail. Eventually, it can identify a cat in a new image it has never seen before — not because it understands what a cat is, but because it has seen those patterns often enough to make an educated guess.

But volume alone isn't enough. Quality matters just as much. Raw data is messy. It may contain missing values, duplicate entries, incorrect labels, or irrelevant information. Even the smartest algorithm can't learn well from disorganized or flawed data.

That's why developers spend a significant amount of time on data preparation:

- **Clean**: Remove duplicates, correct errors, and handle missing values.
- **Transform**: Convert data into formats the algorithm can understand.
- **Normalize**: Ensure numbers are on a similar scale, so the model doesn't misinterpret differences.

In AI there's a simple rule: **"Garbage in, garbage out."** If the input data is biased, inaccurate, or incomplete, the AI will reflect those flaws in its results. It won't know it's wrong — it will confidently repeat the patterns it has been shown.

In short, **data is to AI what experience is to people**. It's how machines build knowledge, spot trends, and make decisions. And just like human

experience, the quality of that experience is what determines how useful, fair, or accurate the AI ultimately becomes.

---

2. How Machines Learn: Three Core Approaches

Artificial intelligence doesn't come preloaded with intelligence. It has to learn — and the process by which it does that is called machine learning (ML). ML is the branch of AI that allows systems to recognize patterns, make predictions, and improve over time by learning from data rather than relying only on hard-coded instructions.

Machine learning powers most of the AI tools we use today — from product recommendations to voice assistants to fraud detection. It isn't one single method; it's a family of approaches designed for different kinds of problems. Just as people can learn by example, by discovery, or by trial and error, machines also have multiple paths to learning.

The three foundational types of machine learning are **supervised learning**, **unsupervised learning**, and **reinforcement learning**. Each method teaches the system in a different way and is suited to a different kind of task.

Supervised Learning

Supervised learning is the most common and well-understood form. Here, the algorithm is trained using labeled data — meaning the inputs come with the correct answers already attached. It's like learning with a teacher who always shows the solution, allowing the AI to figure out the relationship between inputs and outputs.

Take spam detection. The system is shown thousands of emails, each labeled "spam" or "not spam." Over time, it recognizes patterns — specific words, phrases, links, or sender addresses that are more likely to appear in spam. Once trained, the model can evaluate new, unlabeled emails and predict whether they belong in the inbox or the junk folder.

Supervised learning is especially useful when historical examples exist and the goal is to predict or classify something.

Used for: classification, prediction, pattern recognition

Examples: spam filters, credit-risk scores, image classification

Unsupervised Learning

Unsupervised learning is more open-ended. The AI is given **unlabeled data** — there are no predefined answers. Its job is to explore the data and discover hidden patterns, groupings, or relationships. It's like trying to solve a puzzle without the picture on the box.

A common use case is **customer segmentation**. A marketing team might feed an algorithm raw purchase data from thousands of users. Without any prior labels, the AI discovers clusters of similar behavior — maybe "budget shoppers," "loyal regulars," and "holiday splurgers." Those insights can then be used to tailor campaigns or offers.

Unsupervised learning is powerful for uncovering structure in complex datasets, especially when you don't yet know what you're looking for.

Used for: clustering, anomaly detection, data exploration

Examples: grouping customers, detecting unusual transactions, compressing data

Reinforcement Learning

Reinforcement learning is about learning by doing. The AI interacts with an environment, makes decisions, and gets feedback in the form of **rewards or penalties**. It's like training a dog with treats and corrections: good actions are encouraged, bad ones are discouraged. Over time, the system figures out which actions lead to better outcomes.

Imagine training a robot to walk. At first, its movements are random. But every time it takes a stable step, it receives a reward. Eventually, it learns which sequence of movements leads to balance and progress — and which ones lead to a fall.

Reinforcement learning is especially useful when decision-making happens over time and each step affects what happens next.

Used for: robotics, control systems, autonomous navigation, game-playing AI

Examples: self-driving policies, warehouse robots, systems that learn to play chess or Go

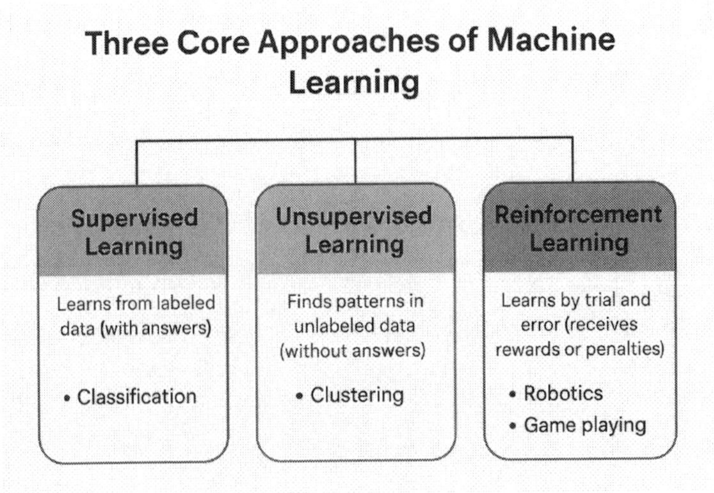

When Learning Goes Wrong: Overfitting and Underfitting

Learning isn't always perfect. Sometimes AI models learn **too much** or **too little** from their training data:

- **Overfitting:** The model memorizes its examples instead of spotting general patterns. It performs well on training data but fails on new, unseen data.
- **Underfitting:** The model is too simple. It misses important relationships in the data and performs poorly even on the training set.

A good model finds the balance: detailed enough to capture real structure, simple enough to generalize beyond the examples it has seen.

Checking the Model: Accuracy, Precision, and Recall

To make sure an AI model is working well, developers evaluate it with metrics:

- **Accuracy:** How often does the model get the answer right overall?
- **Precision:** Of the cases the model flagged as positive (e.g., "fraud"), how many were truly positive?
- **Recall:** Of all the true positive cases that existed, how many did the model successfully catch?

In high-stakes settings — like fraud detection or medical screening — precision and recall can matter more than overall accuracy. These aren't just academic labels; they guide how systems are tuned, approved, and monitored before they're deployed in the real world.

---

3. Algorithms: AI's Brain

If data is the fuel of AI, then **algorithms are the engine** — the part that turns raw information into action. An algorithm is simply a set of instructions a computer follows to solve a problem or reach a decision. It's not a mysterious black box; it's more like a very strict recipe: "If this happens, do that."

Sorting laundry is a simple example of an algorithm:

"If the shirt is white, put it in this pile; if it's colored, put it in that pile." AI algorithms do the same thing with data, just on a much larger scale and at incredible speed.

Different algorithms suit different problems:

- **Linear regression** fits a line through data points to make predictions.
- **Decision trees** break decisions into step-by-step yes / no questions, making the logic easy to follow.

- **Neural networks** use interconnected layers of "nodes" that adjust their connections as they learn, capturing very complex patterns.

You'll see more about full model architectures later in the book, where we unpack how complete AI systems are designed and deployed.

Training vs. Inference

AI systems operate in two main phases:

- **Training:** The model learns from large datasets, adjusting its internal parameters to fit the patterns in the data.
- **Inference:** The trained model is used to make predictions or generate outputs on new data.

Training is like school; inference is like using what you learned on the job.

All algorithms — simple or complex — share the same goal: make sense of data and produce useful output. They don't think or feel. They follow rules and detect patterns. But with the right data and design, those patterns can appear surprisingly intelligent.

---

4. Pattern Recognition: AI's Superpower

AI doesn't reason or imagine the way humans do. Its superpower is **pattern recognition** — finding structure in data that would be hard or impossible for a person to see.

Show a computer thousands of images of handwritten numbers, and over time it learns the features that distinguish a "2" from a "5" or a "7" from a "1." It's not memorizing individual images; it's learning the visual patterns that define each digit so it can guess correctly when it sees new ones.

That same principle shows up everywhere:

- **In marketing**, AI learns what a customer who's about to cancel a subscription "looks like," based on behavior patterns.
- **In finance**, it spots unusual spending patterns that may indicate fraud.
- **In healthcare**, it identifies early signs of disease in medical scans by comparing thousands of cases and finding shared characteristics.

What makes AI powerful here is **scale**. It can process millions of data points in seconds and uncover patterns we'd never notice. But it's not infallible. If the data is messy, biased, or incomplete, the patterns it learns may be misleading or unfair. That's why clean, representative data — and human oversight — still matter.

At its core, **pattern recognition** is what makes AI feel intelligent. It's how machines find order in chaos and make sense of the world without ever truly understanding it.

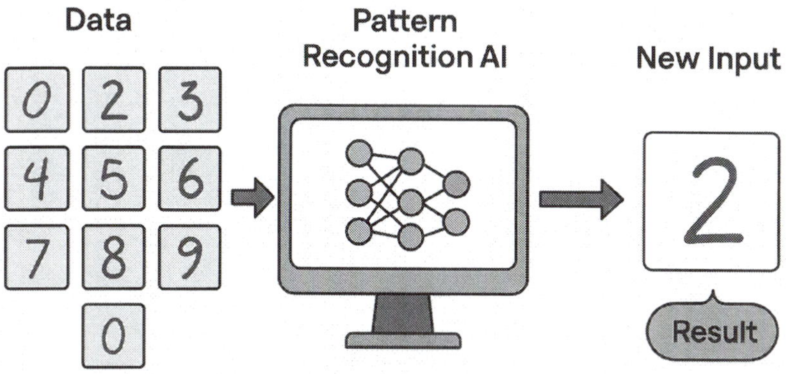

---

5. The Feedback Loop: Learning from Mistakes

AI doesn't have to stay static. One of its most powerful features is the **feedback loop** — a cycle that lets it improve over time.

Imagine a self-driving car that misidentifies a stop sign because it's covered in graffiti. Engineers capture that failure and add new examples of defaced stop signs to the training data. The model is retrained, updates its understanding, and gets better at recognizing tricky signs in the future. The mistake becomes a lesson.

Feedback can work in different ways:

- **Batch updates:** Developers periodically retrain the model on new data collected over weeks or months.
- **Online learning or reinforcement learning:** The system adjusts more continuously as it interacts with the environment and receives rewards or penalties.

Either way, feedback turns a one-off model into a dynamic system that can adapt to new inputs, changing environments, and unexpected situations.

But feedback isn't always automatic. Many AI systems still depend on humans to notice errors, correct outputs, and decide when a model needs retraining. Without that oversight, an AI might keep making the same mistake — and no one would know.

In short, **feedback** is what makes AI systems less brittle and more resilient. They don't grow by understanding, but by **adjusting**.

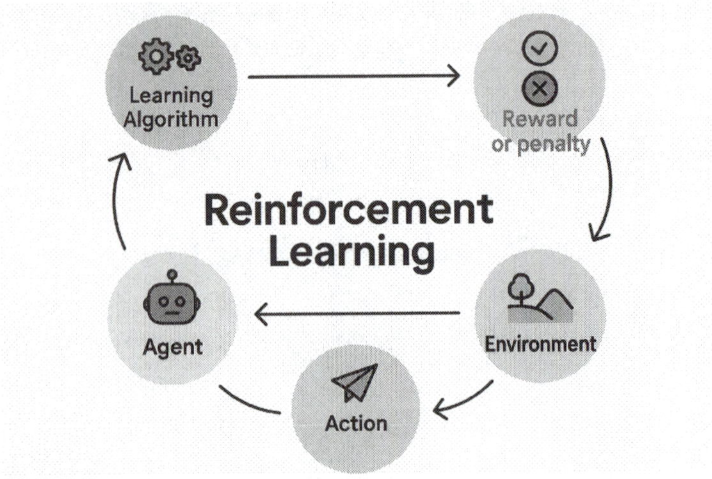

6. Human-in-the-Loop (HITL): Why Humans Still Matter

Even the smartest AI systems make mistakes — and some decisions are too important to leave entirely to machines. That's where **Human-in-the-Loop (HITL)** systems come in.

HITL refers to any AI setup where humans are intentionally kept in the decision process. Sometimes that means reviewing and correcting the AI's outputs. Other times, it means training the system by labeling data, tweaking model behavior, or deciding when automation should take over and when it should stop.

Examples include:

- In **healthcare** or **law**, AI can help flag risks or suggest options, but a human expert makes the final call.
- In **customer service**, an AI might handle simple questions but escalate complex or sensitive cases to a person.
- During **model training**, humans label examples, verify outputs, and step in when the system goes off track.

HITL is essential for:

- **Safety** in high-risk environments (aviation, medical diagnosis, critical infrastructure)
- **Fairness** when decisions involve values, ethics, or social impact
- **Accountability**, ensuring a person — not a model — is ultimately responsible
- **Better learning**, by providing high-quality feedback and corrections

Keeping humans in the loop doesn't slow progress; it makes progress safer, more trustworthy, and more aligned with human goals.

---

7. Limitations of AI: What Machines Still Can't Do

As powerful as AI is, it has clear boundaries.

Most AI today is **narrow**: it's trained for a specific task in a specific domain. It performs best when the problem is well-defined and the data is clean. It struggles with ambiguity, nuance, or decisions that require common sense or moral judgment.

AI also has limited **generalization**. A model trained to detect tumors on medical images can't suddenly read tax returns or mediate a conflict. It can't transfer its knowledge between tasks the way humans routinely do.

Then there's **explainability**. With some systems — especially large neural networks — even the developers can't always say exactly why a particular decision was made. This "black box" problem raises serious questions in areas like healthcare, finance, and criminal justice, where reasons matter.

In short:

- AI doesn't understand meaning.
- It doesn't have intentions or awareness.
- It doesn't ask "Should we?" — only "What's the pattern?"

What AI does is powerful: it automates perception, finds structure in data, and helps us make better decisions. But it's not magic — and knowing its limits is just as important as knowing its strengths.

Quick Takeaways: How AI Thinks

» AI doesn't understand — it detects patterns.

It finds structure in data and uses it to make decisions, predictions, or generate outputs.

» Machines learn in different ways.

Supervised (with answers), unsupervised (without answers), and reinforcement (through trial and error) are the three main approaches.

» Algorithms are the core logic.

They're like recipes that process inputs and deliver results — from simple trees to deep networks.

» Data quality is everything.

Clean, complete, unbiased data leads to better models. "Garbage in, garbage out" applies.

» Feedback powers improvement.

AI learns by adjusting itself over time, using past outcomes to improve future results.

» Humans still belong in the loop.

For safety, fairness, and accountability, people remain essential — especially in high-stakes decisions.

Chapter Summary

AI doesn't think like a person. It doesn't feel, imagine, or understand meaning. What it does do is learn from data — finding patterns, following algorithms, and improving through feedback.

This chapter unpacked what's really going on under the hood:

- We began with **data** as AI's experience — messy at first, but cleaned, transformed, and normalized so models can learn from it.
- We explored **machine learning**, including supervised, unsupervised, and reinforcement learning, and saw how models can overfit, underfit, and be evaluated with metrics like accuracy, precision, and recall.
- We looked at **algorithms** and **pattern recognition**, from simple rules to deep neural networks that uncover complex structure in images, language, and behavior.
- We examined **feedback loops** and **human-in-the-loop** systems, which help AI adapt, stay safe, and remain aligned with human goals.
- Finally, we highlighted the **limits of AI** — its narrow focus, lack of real understanding, and challenges around explainability.

In the end, AI "thinks" by processing examples, detecting patterns, and learning from outcomes — not by reasoning like a human, but by modeling what it has seen. Understanding that difference is key to using AI wisely in the chapters, tools, and decisions that follow.

Part II Inside AI — Core Concepts in Depth

Introduction

In Part 1, we laid the foundation: what AI is, how it differs from human intelligence, where it shows up in everyday life, and how it "thinks" using algorithms, data, and learning. Now it's time to open the black box.

Part 2 shifts from what AI does to how it actually works under the hood. You'll see how modern systems are built, trained, and improved — and what terms like *"deep learning"*, *"foundation models,"* or *"NLP"* really mean in practice.

Rather than offering a dry glossary, this section follows the natural life cycle of an AI system: from core model design to the ability to generate creative content, understand language and images, adapt over time, and make decisions with humans still in charge. Along the way, you'll also see why trust, fairness, and explainability matter just as much as performance.

A useful way to picture modern AI is as a **factory**:

- **Data** is the raw material arriving at the loading dock — messy, inconsistent, and often unstructured.
- **Algorithms** are the machinery that processes that raw material — sorting, filtering, shaping, and refining it.
- **Machine learning models** are the assembly lines — trained through repetition, adjusting to produce increasingly accurate outputs.
- **Deep learning** adds automation and complexity — like robotic arms and smart sensors spotting patterns humans might miss.
- **Foundation models** are massive, multipurpose factories — trained once on huge amounts of data and then reused across many tasks, from summarizing documents to generating code.
- And throughout the factory, **humans** still play key roles — as designers, quality-control inspectors, and safety monitors.

This part isn't about turning you into an engineer. It's about giving you a clear mental model of how modern AI systems are put together, so that when someone mentions tokens, weights, fine-tuning, or model—drift, you'll know what's at stake and why it matters.

By the end of Part 2, you'll have a much clearer sense of how today's most advanced AI systems actually work — and where their power and their limitations come from.

Chapter 8

WHAT MAKES AN AI SYSTEM WORK?
HOW DATA, MODELS, AND ALGORITHMS COME TOGETHER

Introduction: Building Intelligence from the Ground Up

By now, you know AI doesn't think like a human — it doesn't reason, imagine, or understand. What it does do is learn from patterns in data and apply those patterns to new situations. But what actually powers that learning under the hood?

What turns millions of examples into something that can spot tumors, write essays, translate languages, or flag fraudulent transactions?

In this chapter, we unpack how AI systems are built — not as isolated parts, but as a coordinated stack:

- raw data
- algorithms that extract patterns
- models that learn
- systems that use those models in the real world

Think of this chapter as a guided tour behind the curtain. You'll see how an AI system is trained, deployed, and improved — and why the pipeline matters just as much as the model itself.

---

1. The AI Stack: Data → Model → Output

Think of an AI system like a factory:

- **Data** is the raw material arriving at the loading dock.
- **The model** is the machine that transforms it.
- **The output** is the finished product — a decision, prediction, or response.

None of this happens automatically. Like any factory, an AI system needs:

- design (architecture)
- machinery (algorithms)
- training (repetition to improve output)
- evaluation (quality checks)
- human oversight (safety and accountability)

They don't just "run." They're built, optimized, tested — and constantly improved.

This pipeline is often referred to as the AI stack, and the simplified version looks like this:

1. **Data** — collected, cleaned, labeled, and prepared
2. **Algorithm** — a method for learning patterns
3. **Model** — the trained "machine" that performs the task
4. **Inference** — using the model on new, unseen data
5. **Output** — a prediction, classification, or generated response

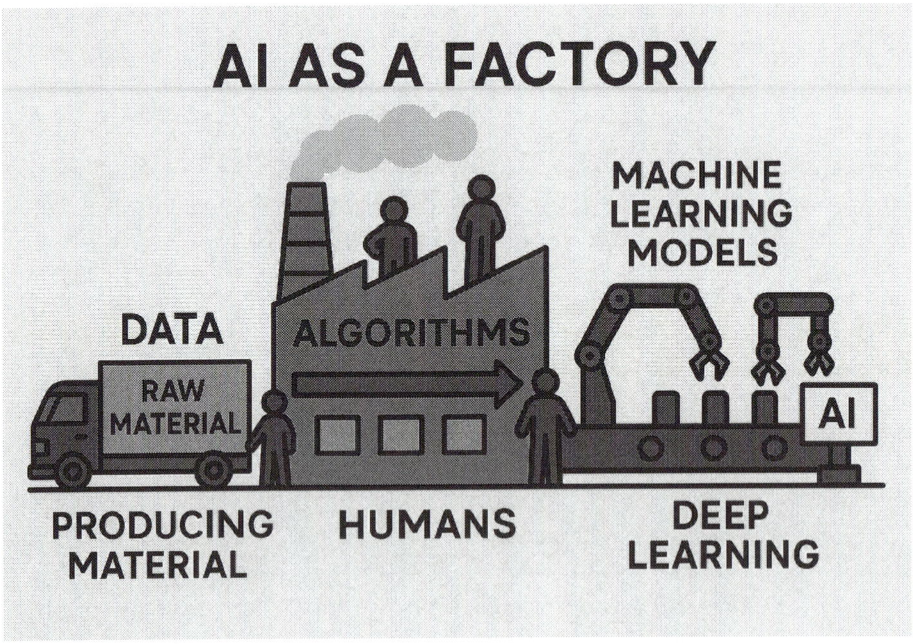

Let's unpack each of those components in more depth.

2. Machine Learning in Context: How the "Learning" Happens

Machine Learning (ML) is the engine that connects data to decisions.

You've already met the three main types — supervised, unsupervised, and reinforcement learning — so here we keep the focus practical:

What is ML doing during training?

It is **adjusting internal parameters** in response to examples until the model can make accurate predictions on its own. These parameters — often called weights — are what the algorithm adjusts during training to reduce error.

You can think of ML as:

» **Algorithms that gradually tune a model to recognize patterns in data.**

The "machine" isn't the computer — it's the learned model.

---

3. Models vs. Algorithms vs. Systems

These words get confused constantly. Here's the clean distinction:

Algorithm

A method or procedure for learning from data.

Examples: gradient descent, decision trees, backpropagation.

Model

The result of applying an algorithm to data — the trained structure that can make predictions.

System

Everything surrounding the model:

- data pipelines
- serving infrastructure
- monitoring tools
- interfaces
- human oversight
- feedback loops

In short:

» **Algorithms train models. Models power systems.**

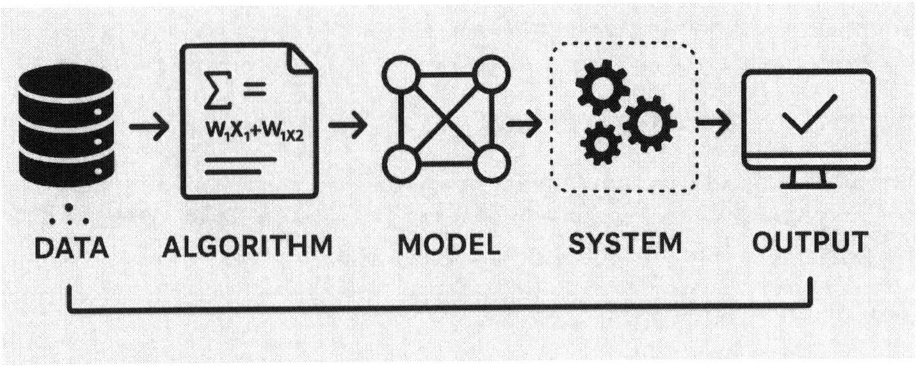

---

4. Training vs. Inference: The Two Lives of an AI System

AI has two distinct phases:

Training (Learning Phase)

- Takes place behind the scenes
- Uses massive datasets
- Adjusts internal weights to reduce errors
- Is slow and computationally expensive
- Often requires specialized hardware

Example:
A vision model might train for weeks on millions of medical images.

Inference (Doing Phase)

- Happens after deployment
- Uses what the model learned to make real-time predictions
- Is fast and optimized for everyday use

Example:
That same model can evaluate a new X-ray in milliseconds.

A helpful analogy:

» **Training is like going to school.**

» **Inference is using what you learned in real life.**

5. Model Architectures: Different Tools for Different Jobs

Models come in many forms, each suited to a type of problem.

Linear Models

The simplest models — perfect for problems where relationships are clear.

Used for:

- price prediction
- basic classification
- risk scoring

Strengths: simple, fast, transparent
Limitations: can't capture complex patterns

Decision Trees & Ensemble Models

Decision trees break decisions into branches.

When you combine many of them — random forests, gradient boosting — you get powerful ensemble models.

Used for:

- fraud detection
- credit scoring
- tabular business data

Strengths: interpretable, strong performance

Limitations: limited on unstructured data like images or text

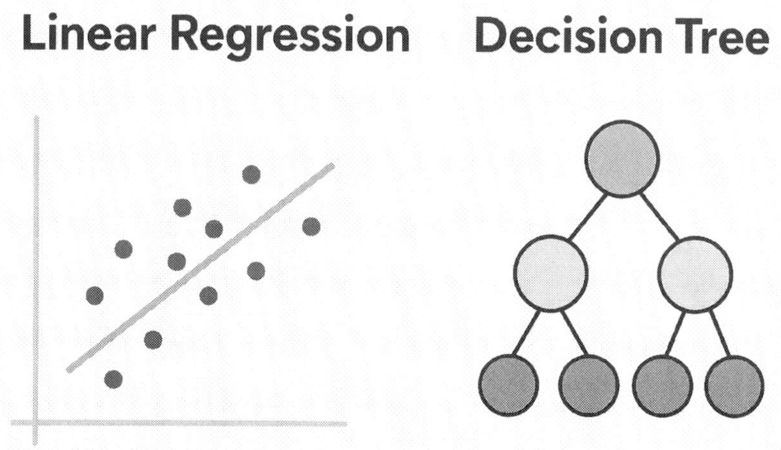

Neural Networks

Neural networks are the backbone of most modern AI systems. They're loosely inspired by the structure of the human brain — but in reality, they're just layers of math, not biological neurons.

At a high level, a neural network is a system of **virtual "neurons"** arranged in layers. Each neuron receives numbers as input, performs a simple calculation, and passes a result to the next layer. What makes this powerful is:

- There are many neurons
- They're organized in multiple layers
- The connections between them have **weights** that can be adjusted

Those weights determine how strongly one neuron influences another. During training, the network repeatedly adjusts those weights so that it becomes better and better at a task.

A basic neural network has three kinds of layers:

Input layer

Where data first enters the network.

- For images: pixel values
- For tabular data: numbers like age, income, temperature
- For text: numerical representations of words or tokens

Hidden layers

Where most of the work happens. Each hidden layer transforms the input a little more, detecting patterns and combinations that aren't obvious in the raw data.

- Early layers might detect simple features (edges, colors, basic shapes)
- Deeper layers pick up more abstract features (faces, objects, sentiment, topics)

Output layer

Produces the final result.

- "Cat" vs "dog"
- "Spam" vs "not spam"
- A number (like a predicted price or probability)

Forward Propagation and Backpropagation

A neural network learns through a two-step loop:

1. **Forward propagation**
    - Input data flows through the network layer by layer
    - Each neuron performs a calculation and passes its result on
    - The network produces an output — essentially, its best guess
2. **Backpropagation**
    - The output is compared to the correct answer
    - The difference is called the error or loss
    - That error is sent backwards through the network
    - The network adjusts the weights of its connections to reduce future error

This process repeats thousands or even millions of times across many examples. Over time, the network becomes very good at mapping inputs to outputs — for example, recognizing a tumor in a scan or identifying spoken words in noisy audio.

The more high-quality data a neural network sees, the better it tends to perform. That's why neural networks are especially powerful in domains with large, rich datasets such as images, text, audio, and sensor data.

Different Types of Neural Networks

Different problems call for different neural network architectures:

Perceptron

The simplest form — a single layer connecting inputs directly to outputs. It's mostly of historical interest now, but it's the conceptual ancestor of more complex networks.

Feedforward Neural Networks

Data flows in one direction: from input to output, with one or more hidden layers in between. These are used for many classification and regression tasks.

Deep Neural Networks (DNNs)

Feedforward networks with **many** hidden layers. The term "deep learning" comes from this depth. More layers allow the network to learn more complex, layered patterns.

Modular Neural Networks

Multiple smaller networks work on different parts of a problem, and their outputs are combined. This can make systems more flexible and scalable.

Convolutional Neural Networks (CNNs)

Designed for visual data.

- Use small "filters" that slide across an image
- Detect local patterns like edges, corners, textures
- Combine these into higher-level features (eyes, wheels, buildings)

Recurrent Neural Networks (RNNs) and related models

Designed for sequences: language, time series, audio.

- Retain a form of "memory" of previous inputs
- Useful for predicting the next word in a sentence or the next value in a series

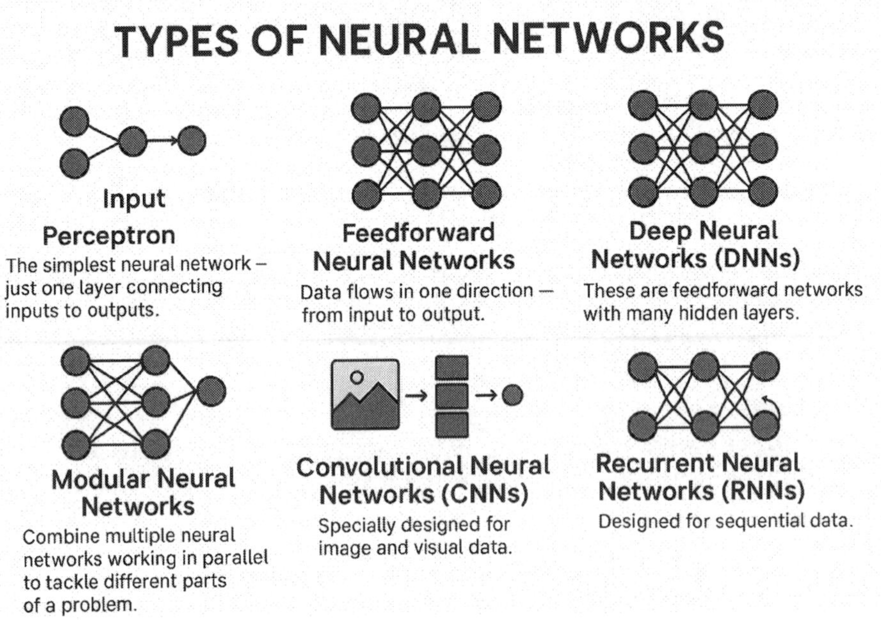

Today, many of the most powerful systems (like transformers) build on or extend these ideas — but the core concept remains: **data flows through layers, connections are weighted, and those weights are learned.** (We'll explore transformer models in depth in Chapter 11.)

Neural networks power everything from face recognition and voice assistants to fraud detection and medical diagnosis. They don't "understand" in a human sense, but they are extraordinarily good at recognizing patterns once trained on enough data.

Used for: image recognition, language modeling, audio processing, fraud detection, medical imaging, time-series analysis.

Deep Learning

Deep learning is a specialized branch of machine learning built on **neural networks with many layers**. Where traditional models might use one or two layers, deep learning models can have dozens or even hundreds.

This depth gives them two major advantages:

- They can work directly with **unstructured data** like raw images, audio, or text
- They can **discover features automatically**, instead of relying on humans to specify which patterns to look for

How Deep Learning Works (Conceptually)

Deep learning models process data step by step:

1. **Input**

   The model receives raw or lightly processed data — for example, an image, a sentence, or an audio clip.

2. **Layered transformations**

   Each layer extracts slightly more abstract features.

   In an image model:

   - Early layers detect edges and simple textures
   - Middle layers detect parts like eyes, wheels, leaves
   - Later layers detect full objects like faces, cars, or trees

3. **Prediction**

   The final layer combines all the learned features and outputs a prediction: "dog," "spam," "benign," "positive sentiment," and so on.

4. **Learning through error correction**

   During training, the model compares its predictions with the correct answers, computes an error, and adjusts its internal weights to reduce that error next time (using backpropagation).

   This process repeats across massive datasets, gradually refining the model until it becomes highly accurate.

Why Deep Learning Became a Breakthrough

Deep learning scales in a way older methods couldn't:

More data → better performance

Older models often plateau after a certain amount of data. Deep learning models tend to keep improving.

More compute → more complex patterns

Deep learning requires significant computational power — this is also why it depends heavily on GPUs and cloud-scale infrastructure.

With powerful GPUs and cloud infrastructure, deep networks can tackle problems that were previously out of reach (like real-time translation, self-driving perception, or large-scale language generation).

That's why deep learning underpins many of the AI systems that feel most "intelligent" today, including:

- speech recognition and real-time transcription
- facial recognition and emotion detection
- medical image diagnosis
- language translation and large language models
- autonomous vehicles and robotics

They still don't "understand" in a human way — but they can capture intricate patterns in our data and make remarkably good predictions based on them.

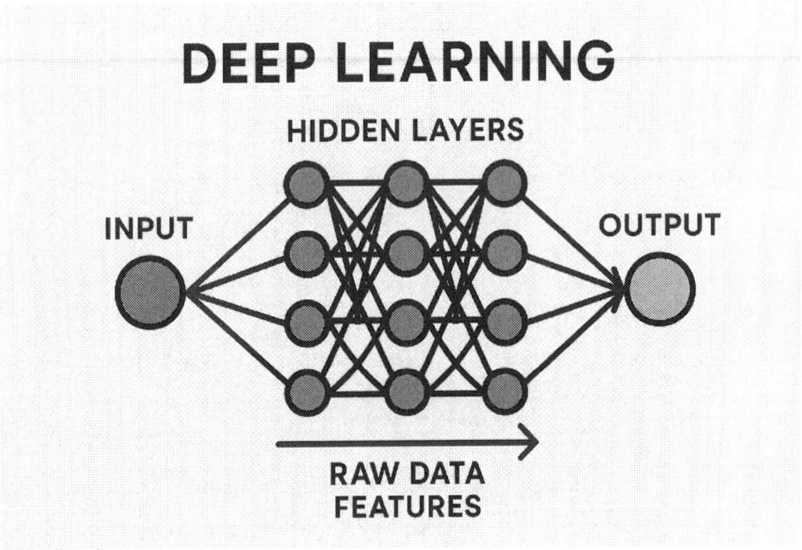

---

6. Machine Learning vs. Deep Learning

Machine learning (ML) and deep learning (DL) share the same goal — learning from data — but they differ in how they get there and what they're best at.

The Hierarchy

Artificial Intelligence (AI)

The broad field of building machines that perform tasks we associate with intelligence.

Machine Learning (ML)

A subfield of AI where systems learn patterns from data rather than being fully hand-programmed.

Deep Learning (DL)

A subfield of machine learning that uses deep neural networks to learn complex patterns, often from unstructured data.

Classical Machine Learning: Structured and Human-Guided

In traditional ML, humans do a lot of upfront thinking:

- They decide **which inputs (features)** matter
- They prepare **structured datasets** (tables, labeled examples)
- They choose models like decision trees, logistic regression, or support vector machines

Example: deciding what shoes to recommend.

You might feed the model structured data like:

- Weather: rainy (1) or not (0)
- Activity: formal (1) or casual (0)
- Distance: long walk (1) or short trip (0)

You then train a model that learns how these inputs relate to different shoe choices.

Characteristics of classical ML:

- Works best with **structured, well-defined data**
- Requires **manual feature engineering** (humans decide what to feed the model)
- Often easier to **interpret and explain**

Deep Learning: Unstructured and Data-Driven

Deep learning changes the workflow:

- You feed the model **raw or lightly processed data**

  Images, audio, text, sensor readings

- You don't have to hand-design as many features
- The network **discovers patterns and features** automatically

Back to the shoe example: instead of structured 0/1 flags, you might give the system:

- Outfit photos
- GPS context
- Free-text descriptions like "important meeting" or "going hiking"

The deep learning model learns which combinations of cues correspond to which shoe choices — without you specifying all the rules.

Characteristics of deep learning:

- Excels with **unstructured or high-dimensional data**
- Learns **features automatically** from raw inputs
- Requires **more data and compute**, but can outperform classical ML on complex tasks

Side-by-Side Comparison

	Machine Learning	Deep Learning
Data type	Structured (tables, numeric fields)	Unstructured (images, audio, text, video)
Feature design	Human-designed features	Features learned automatically
Architecture depth	Shallow (few layers)	Deep (many layers of neurons)
Data requirement	Works with smaller datasets	Thrives on large datasets
Interpretability	Often easier to explain	Often more opaque ("black box")
Typical use cases	Credit scoring, churn prediction, tabular BI	Vision, speech, language, complex pattern tasks

Both approaches are valuable. In practice:

- If you have **clean tables and limited data**, classical ML can be ideal.
- If you're working with **images, audio, or natural language at scale**, deep learning is usually the better fit.

---

7. Wrapping Up: Behind the Curtain of AI

AI systems can look mysterious from the outside, but they're built from understandable parts:

- **Data** — the examples they learn from
- **Algorithms** — the procedures that adjust the model
- **Models** — the structures that encode what's been learned
- **Systems** — everything that wraps around the model to make it useful
- **Feedback loops** — how performance is monitored and improved over time

Neural networks and deep learning don't "think" or "understand" — they **optimize**. They turn vast amounts of data into patterns, and patterns into predictions or generated content.

Whether it's a simple linear model or a deep network with billions of parameters, the goal is the same: turn raw input into **useful output**. The better you understand that process, the more confidently you can evaluate AI systems, ask the right questions, and decide when — and when not — to trust their results.

Quick Takeaways: What Makes an AI System Work

» **AI systems are built, not born.**

They rely on structured components — data, algorithms, models, and feedback loops — that work together to turn information into predictions or actions.

» **Machine learning connects data to decisions.**

ML trains models to recognize patterns from examples and generalize to new situations.

» **Training is learning; inference is doing.**

Training adjusts the model's internal parameters; inference uses the trained model to make predictions in real time.

» **Neural networks pass information through layers.**

Each layer transforms the data, and the model learns by adjusting connections to reduce error.

» **Deep learning uses many layers to learn features automatically.**

It handles images, audio, language, and other messy data by discovering patterns without human-defined rules.

» **Machine learning and deep learning differ in complexity and data needs.**

ML works best with structured data and human-crafted features; DL excels with unstructured data but needs far more data and compute.

Chapter Summary

In this chapter, we pulled back the curtain on how AI systems actually work. We began with the AI stack — data in, models trained, outputs generated — and explored how these components turn raw information into intelligent behavior.

We clarified the roles of **algorithms** (the learning procedure), **models** (the trained result), and **systems** (the full pipeline that makes AI operational). We explained how training and inference differ, and why most of the heavy computation happens before a model is ever deployed.

We then walked through the main families of models — from linear regressions and decision trees to deep neural networks — and took a deeper look at how neural networks learn through forward propagation, backpropagation, and weight adjustment. Special attention was given to deep learning and why it excels at tasks like image recognition, speech processing, and language understanding.

Finally, we compared machine learning and deep learning, highlighting the differences in data needs, feature engineering, interpretability, and where each approach shines.

The big takeaway:

AI systems aren't magical. They're engineered tools that identify patterns, optimize based on feedback, and generate predictions — powerful, but not mysterious. Understanding how they work helps us use them more wisely, responsibly, and effectively.

Chapter 9

DEEP LEARNING AND FOUNDATION MODELS

Introduction: From Deep Learning to Foundation Models

In the last chapter, we explored how AI systems work — from the data and algorithms that power them, to the neural networks and deep learning models that make them flexible and powerful. This chapter picks up where we left off, zooming in on how deep learning evolved into something even bigger: **foundation models**.

These models don't just perform one task. They're trained on massive amounts of data and designed to handle a wide range of activities — summarizing, translating, reasoning, answering questions — all with the same underlying engine. They've reshaped what AI can do.

But how did we get here? What makes these systems so capable? And what are their limits?

---

1. What Makes Deep Learning Different?

Traditional machine learning works well when the data is structured and the task is narrow. But it starts to break down when problems get messy — when the data is unstructured, the context is complex, or the relationships are too subtle to program manually.

Deep learning changes that. It doesn't need hand-crafted features or rigid categories. It can learn directly from raw data — text, images, audio — and extract what matters on its own. The key is scale: more layers, more data, more compute.

This ability to learn rich, hierarchical representations of data is what allows deep learning models to generalize better — and unlock tasks that once seemed out of reach.

2. Neural Networks → Deep Learning → Transformers

Transformers changed everything. Introduced in 2017, the Transformer architecture replaced earlier models that processed sequences step-by-step (like RNNs and LSTMs) with a structure that processes entire sequences in parallel. The breakthrough was a mechanism called **attention**, which allows the model to decide which words or tokens to focus on — regardless of their position in the text.

This attention-based approach solved two major bottlenecks:

- older models struggled with long-range dependencies
- sequential processing limited speed and scalability

Transformers avoid both issues.

Originally, transformers had two components:

- **Encoder** — processes and understands the input
- **Decoder** — generates output autoregressively

Modern large language models used in today's AI assistants typically rely on **decoder-only transformers**, which are optimized for generating text. **Encoder-only transformers**, such as the family introduced by Google for language understanding, still excel at classification and comprehension tasks.

Why Transformers Matter:

- They capture long-range relationships extremely well.
- They scale efficiently with compute and data.
- They support pretraining → fine-tuning → generalization across many tasks.
- They form the backbone of today's most capable LLMs used in chatbots, search, translation, and analysis.

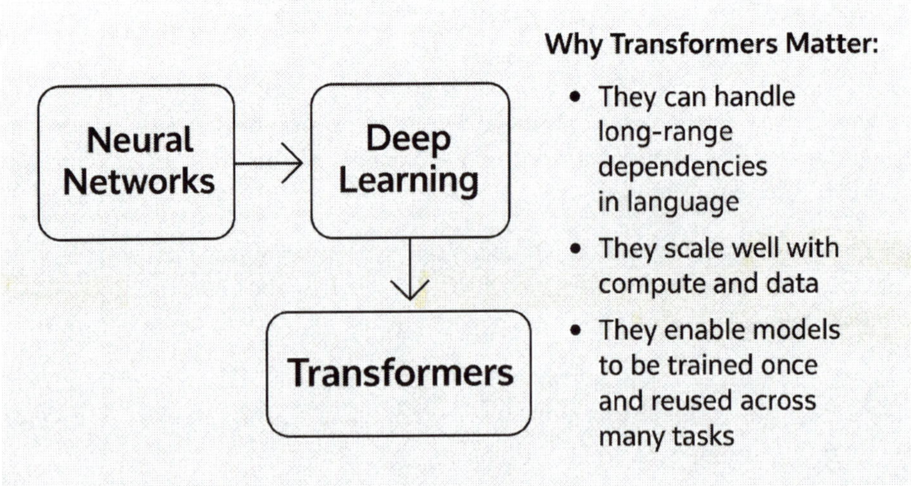

3. Enter the Foundation Model: Pretrained Giants

The term "foundation model," introduced by Stanford's Center for Research on Foundation Models (CRFM), describes a shift in how AI systems are built and deployed. Historically, every AI task had its own model trained from scratch — image classification, spam detection, translation, sentiment analysis, and so on.

Foundation models broke that pattern.

Instead of training many small specialized models, we train **one extremely large, general-purpose model** on diverse datasets (books, web text, code, images, etc.) and then adapt it for downstream tasks. The same pretrained model becomes the starting point — the foundation — for dozens or even thousands of applications.

What makes it a "foundation" isn't only scale. It's that:

- it is trained once (using huge compute)
- it acquires broad, cross-domain knowledge
- it can be adapted for new tasks cheaply and quickly
- its representations transfer surprisingly well across domains

This isn't just efficient — it's revolutionary.

Foundation models learn rich internal representations that allow them to:

- classify spam
- summarize documents
- generate code
- answer questions
- translate languages
- reason across loosely connected concepts

All with the same underlying model.

Key Characteristics of Foundation Models:

- Trained on massive, heterogeneous datasets
- Contain billions to trillions of parameters
- Use self-supervised learning (predicting masked words, next tokens, or image patches)
- Require large-scale compute (GPU clusters, TPU pods, specialized accelerators)
- Can be fine-tuned, adapted, or prompted for new tasks

Examples:

- GPT / ChatGPT — OpenAI
- Gemini — Google DeepMind
- Claude — Anthropic
- LLaMA — Meta

These models now serve as the infrastructure layer of modern AI.

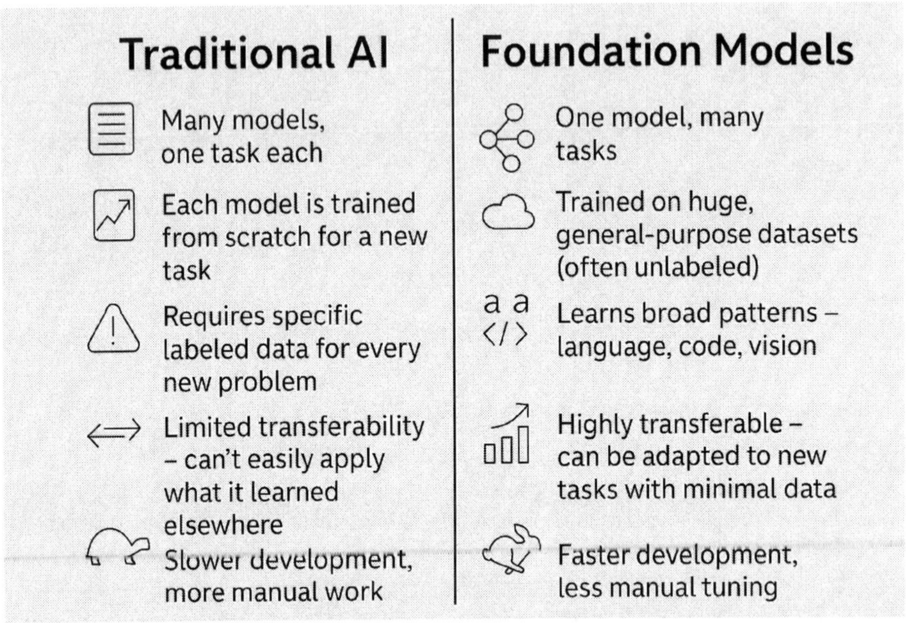

---

4. Fine-Tuning and Customization: From Generic to Specific

Foundation models are powerful, but not perfect out of the box. To make them truly useful in specific contexts — medicine, law, education, customer support — they often need to be adapted. Fortunately, there are several ways to do this without retraining a model from scratch.

These approaches vary in complexity, cost, and how deeply they modify the model. They let organizations customize behavior, tone, and performance for their needs — whether that's answering legal queries, writing with a brand voice, or staying up-to-date with the latest facts.

Fine-Tuning

This method involves taking a general-purpose model and further training it on a smaller, domain-specific dataset (e.g., legal contracts or product manuals).

It's like taking a broadly educated AI and teaching it the specialized language of law, medicine, or customer service.

Fine-tuning plays a foundational role in AI training and will be explored in more depth in a later chapter — especially in how models evolve over time.

Adapters and LoRA (Low-Rank Adaptation)

Instead of retraining the full model, adapters and LoRA offer plug-in components that "bolt on" new knowledge. They're faster, more modular, and ideal for use cases where you need specialization without touching the base model.

Prompt Engineering and API Tools

Sometimes, you don't need to change the model at all. Instead, you guide it with carefully crafted prompts — framing the input in ways that steer the output. Prompt engineering can achieve surprisingly good results with no model changes, especially when using APIs like OpenAI's or Anthropic's.

Retrieval-Augmented Generation (RAG)

RAG enhances generative models by combining them with a search or retrieval system. When a user asks a question, the model pulls in relevant external documents before generating a response. This grounds the output in factual, current data — ideal for tasks that require accuracy and up-to-date knowledge. (We'll explore RAG in greater depth in Chapter 10.)

Together, these customization strategies allow organizations to turn general-purpose models into tailored, high-performance tools — with varying degrees of technical investment, flexibility, and control.

---

5. The Power and Limits of Scale

Scaling deep learning models has unlocked astonishing capabilities. As models grow larger and see more diverse data, they develop internal representations that generalize surprisingly well across tasks — translation, summarization, coding, reasoning, and more.

But scale is not magic.

What scale can do:

- Strengthen general-purpose understanding
- Improve fluency, coherence, and reasoning-like behavior
- Reduce error rates across many benchmarks

- Enable strong zero-shot and few-shot performance
- Make models adaptable with very little task-specific data

What scale cannot do:

- Guarantee factual accuracy
- Eliminate inherited biases from training data
- Prevent hallucinations
- Produce genuine understanding or reasoning
- Replace expert judgment in specialized or high-stakes domains

Large models are powerful statistical engines—not all-knowing systems. They generate the most likely response, not the objectively correct one. That's why oversight, grounding mechanisms (like RAG), and domain expertise remain essential.

Quick Takeaways: Deep Learning and Foundation Models

» Deep learning learns from unstructured data

Layered neural networks allow models to extract meaning and patterns from raw inputs without explicit programming.

» Transformers revolutionized scalability

This architecture made it possible to train large models efficiently by enabling parallel processing and long-range attention.

» Foundation models are versatile starting points

Trained on broad datasets, these models can be adapted to perform many different tasks with minimal additional data.

» Customization enhances utility

Techniques like fine-tuning, adapters, Retrieval-Augmented Generation (RAG), and prompt engineering help tailor models to specific use cases.

» Scale is powerful but complex

Larger models can do more, but they also demand more resources and introduce new risks like bias, opacity, and hallucination.

Chapter Summary

This chapter traced the path from traditional deep learning to foundation models — the engines behind many of today's most powerful AI systems.

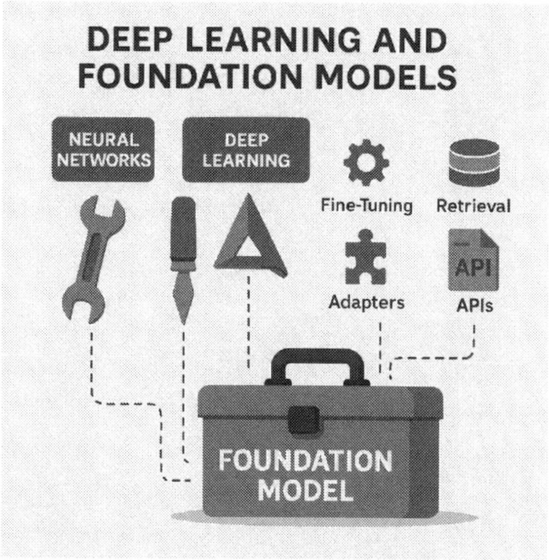

We explored what makes deep learning unique: its ability to learn directly from data, to scale with compute, and to adapt across tasks. We saw how transformer models revolutionized the architecture of AI, and how pretraining large models opened the door to general-purpose AI.

Finally, we looked at how these models are customized for real-world use through fine-tuning, retrieval, adapters, and APIs — and how scale enables both great promise and new challenges.

In short: today's most advanced AI isn't trained for one task — it's trained for many. And with the right tools and care, it can be adapted to do even more.

Chapter 10

RETRIEVAL-AUGMENTED GENERATION (RAG): BRIDGING KNOWLEDGE AND GENERATION

Introduction: Grounding AI in Facts

Foundation models — whether GPT-style or Gemini-style — are incredibly powerful, but they all share one major limitation: **their knowledge is frozen in time**. Once trained, they cannot update themselves, access external databases, or look up new information. This means they can produce answers that sound confident but are outdated, shallow, or simply incorrect — a phenomenon known as hallucination.

Retrieval-Augmented Generation (RAG) solves this problem by pairing a generative model with an external search or retrieval system. Instead of relying solely on what the model "remembers," RAG gives it a live library to pull from. The result is a hybrid architecture: the creativity and fluency of generative AI, grounded in real, factual sources.

---

1. What Is RAG and How Does It Work?

At its simplest, RAG adds one crucial step to the generative process: retrieval.

Instead of generating answers purely from internal parameters, a RAG system:

1. **Retrieves** relevant documents from a database, knowledge base, or curated corpus
2. **Augments** the user's query by adding those documents
3. **Generates** a response using both the query and the supporting information

This "retrieve → augment → generate" loop dramatically improves accuracy. It's like allowing a student to consult reference materials during a test — the answer is still generated creatively, but it's grounded in verifiable facts rather than memory alone.

2. Why Use RAG?

Generative models are powerful but inherently limited by static training data. Once trained, they cannot update their knowledge unless the entire model is retrained — an expensive and resource-intensive process. That leads to predictable problems:

- **Factual inaccuracies** from hallucination
- **Outdated knowledge** as the world changes
- **Shallow answers** when details aren't encoded in the model
- **Limited reasoning depth** due to context window constraints

RAG directly addresses these limitations. By pulling up-to-date, domain-specific, and context-rich documents on demand, it:

- Provides **current** information
- Grounds answers in **real sources**
- Allows deeper and more **contextual** responses
- Supports **long-form reasoning** by retrieving extended documents rather than relying on a short prompt

In short, RAG makes generative models more trustworthy and more useful.

---

3. Core Components of RAG

Retriever

The retriever searches a document store for the most relevant pieces of information. This can be done using:

- Traditional keyword search (e.g., **BM25**)
- Modern vector-based search using embeddings
- Large-scale systems like **FAISS**, **Elasticsearch**, or **BigQuery**

Retrievers determine what information enters the model — which means they heavily influence accuracy.

Generator

Once documents are retrieved, a generative model uses them to produce a final answer. Modern LLMs — whether GPT-style, Claude-style, Gemini-style, or any future architecture — excel at transforming retrieved passages into coherent, fluent, and context-aware responses.

The key is synthesis, not copying.

The generator blends retrieved facts with its own reasoning abilities to produce a tailored answer.

Orchestration Layer

Frameworks such as **LangChain**, **LlamaIndex**, **HuggingFace**, or **Vertex AI** manage the flow:

- Run the retrieval
- Format the augmented prompt
- Pass it to the generator
- Handle outputs and citations

This layer is what turns RAG from a concept into a working system.

---

4. How RAG Improves Generative AI

RAG enhances generative models in four major ways:

1. Grounding
It roots answers in real documents, making results more factual and verifiable.

2. Freshness
It bypasses the model's knowledge cutoff, pulling from up-to-date sources.

3. Contextual Memory
RAG can incorporate long-form documents that exceed the model's token limit.

4. Domain Specificity
By retrieving industry-specific documents — legal, medical, financial, scientific — the model responds with expert-level detail.

Example:
A legal assistant powered by RAG can pull the most relevant case law or internal policies before answering a question — producing an answer that is both fluent and grounded.

---

5. Real-World Applications

RAG is now used across industries, especially in situations where accuracy is critical:

- **Customer support**: grounding answers in policy docs and FAQs
- **Search engines**: combining documents with generative summaries
- **Education**: providing explanations tied to textbooks or academic sources
- **Enterprise tools**: retrieving internal knowledge for decision-making

In high-stakes fields like healthcare or law, grounding answers in verified documents dramatically reduces errors and improves trust.

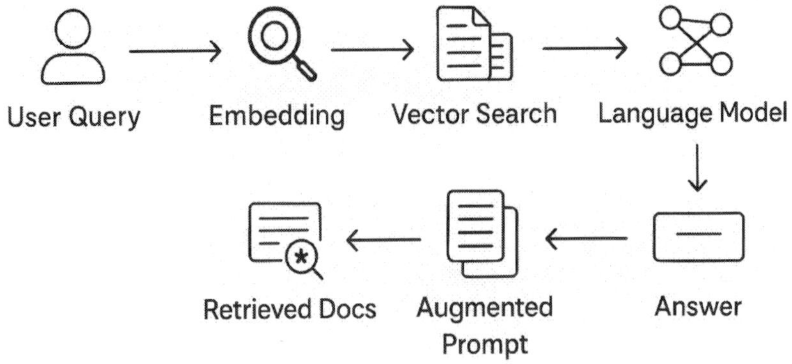

---

6. RAG on Google Cloud

Google Cloud provides a robust environment for building scalable RAG systems:

- **Vertex AI** integrates retrieval with powerful foundation models (PaLM-style or Gemini-style)
- **BigQuery** supports fast, large-scale vector search and document retrieval
- **Embeddings APIs** allow efficient similarity search across millions of documents

A healthcare provider, for example, can retrieve the latest medical studies from BigQuery and have Vertex AI synthesize a patient-friendly explanation — grounded, current, and safe.

---

7. Example in Action

Query:
"What were the key causes of World War II?"

RAG Process:

- Retrieve passages from trusted historical sources
- Add them to the prompt
- Generate a synthesized, historically accurate explanation

The answer is not only coherent — it is verifiable, because the model is grounded in retrieved evidence rather than memory.

Quick Takeaways: Retrieval-Augmented Generation (RAG)

» Grounds generative responses in facts

RAG enhances generative models by retrieving relevant, factual information to support and inform output.

» Reduces hallucinations

By anchoring generated content in real-world documents, RAG helps minimize confident but incorrect answers.

» Overcomes model limitations

RAG supplements static model knowledge, addressing issues like outdated information or lack of domain specificity.

» Works across industries

From healthcare to law to education, RAG can be adapted to provide accurate, context-aware support in specialized fields.

» Scalable with modern tools

Platforms like **Google Cloud Vertex AI** and **BigQuery** offer infrastructure for building, deploying, and scaling RAG-based applications.

Chapter Summary

Retrieval-Augmented Generation bridges a critical gap in modern AI. By pairing retrieval with generation, RAG produces responses that are not only fluent and creative but also accurate, current, and domain-specific. Throughout this chapter, we explored how RAG works, why it matters, and where it's being deployed in the real world.

The takeaway is simple:

Generative AI is powerful, but RAG makes it reliable.

As foundation models continue to evolve, retrieval systems will remain essential for grounding AI in reality — ensuring that the answers it provides are not only impressive but trustworthy.

Chapter 11

GENERATIVE AI — SYSTEMS THAT CREATE

Introduction: From Recognition to Generation

For most of AI's history, systems were built to recognize things — is this a cat or a dog? Is this transaction fraudulent or not? Generative AI changed that trajectory. These models don't just interpret the world; they create new content: text, images, code, music — even video.

This shift from recognition to generation marks a major turning point in what AI can do. Instead of choosing among known options, generative systems produce new possibilities. They draw from patterns in the data they were trained on and generate outputs that are often convincingly humanlike.

In this chapter, we unpack how generative AI works, the types of content it can produce, and the implications — both promising and challenging — of machines that can generate.

---

1. How Generative Models Work

Generative AI systems are trained not just to classify or predict, but to produce new content. Instead of selecting a label or category, they model probability distributions — essentially the patterns of how things tend to appear — and sample from those patterns to create original outputs.

Several core techniques power today's generative AI:

Transformers

The backbone of most modern generative systems. Transformers are deep learning architectures that process data in parallel and use **attention mechanisms** to understand relationships between parts of a sequence — such as words in a sentence or patches in an image.

Used in: ChatGPT, Claude, Gemini, LLaMA, Copilot.

Diffusion Models

Diffusion models generate images by reversing a controlled noise process. They start with pure randomness and gradually "denoise" it into a coherent image — like sculpting a scene out of static.

Used in: DALL·E 2, Midjourney, Stable Diffusion.

GANs (Generative Adversarial Networks)

GANs use two competing networks: a generator that creates images and a discriminator that judges them. Through this competition, the generator improves until it can consistently fool the discriminator.

Used in: Deepfakes, artistic synthesis, synthetic photography.

Each method excels in different areas:

- **Transformers** dominate in language and mixed-modality tasks.
- **Diffusion** models produce high-quality, controllable images.
- **GANs** are fast and sharp but more challenging to train.

---

2. Modalities of Output: What Can Be Generated

Generative AI models are not confined to a single type of output. Unlike earlier AI systems, which might classify an email as spam or identify a cat in a photo, generative systems can create new content across multiple formats. This capability — known as multimodality — allows a single architecture to understand and generate different types of media.

Let's explore the major modalities and how they're being used:

Text: Writing, Summarizing, Translating, and Beyond

Text generation is the most mature and widespread application of generative AI. Models like GPT, Claude, and Gemini can:

- Write emails, blog posts, essays, marketing copy, or poetry
- Summarize long documents, transcripts, or reports
- Translate between dozens of languages with fluency
- Rewrite or rephrase text for tone, clarity, or brevity
- Simulate human conversation in chatbots or virtual assistants

Use cases:

- Customer support
- Content marketing
- Academic research
- Journalism
- Legal document review

What makes it powerful: adaptability — the same model can write a newsletter, explain physics, or craft a bedtime story depending solely on the prompt.

Images: From Concepts to Visual Realities

Generative image models like **DALL·E 2**, **Midjourney**, and **Stable Diffusion** create compelling visuals from text prompts. These systems interpret language and convert it into detailed imagery — often in specific styles or formats.

Types of images generated include:

- Concept art and illustration
- Product mockups and prototypes
- Logos, icons, and branding assets
- Photorealistic or stylized portraits
- Landscapes, characters, and fictional worlds

Used by: designers, marketers, educators, authors, game developers.

These tools **dramatically accelerate** visual ideation — work that once required hours of design can now be prototyped in seconds.

Audio: Voices, Music, Soundscapes

Generative audio has advanced rapidly. AI can now:

- Synthesize realistic voices
- Clone voices for narration or dubbing
- Compose original music
- Generate ambient soundscapes and effects
- Convert text to speech with emotion and nuance

Models like **Udio**, **ElevenLabs**, and **Suno** are pushing the frontier. While challenges remain (e.g., copyright, emotional depth), the quality and control of generative audio are improving quickly.

Code: Writing, Fixing, and Explaining Software

AI coding assistants such as **GitHub Copilot**, **Claude**, and **GPT-4** are transforming how developers work. These tools can:

- Generate code from natural language prompts
- Suggest autocompletions while typing
- Debug, refactor, and optimize code
- Translate code between languages (e.g., Python to JavaScript)
- Explain how complex code works in plain English

This isn't just for experts. AI coding tools are lowering the barrier to entry for beginners and accelerating productivity for pros.

Examples:

- Automating repetitive scripts
- Testing and documentation
- Rapid prototyping for apps and websites

Video: The Next Frontier

Generative video is still emerging but holds immense potential. Early-stage tools like **Runway**, **Pika**, and **Sora** can:

- Generate short clips from text prompts
- Animate still images or characters
- Create motion graphics and transitions
- Replace backgrounds, faces, or voices in post-production
- Simulate virtual actors or scenes from scratch

While currently limited by realism, duration, and coherence, video generation is advancing fast. Future applications may include:

- Virtual news anchors
- Personalized ads
- Film pre-visualization
- Synthetic training simulations

Why Multimodality Matters

What makes these modalities so powerful isn't just that AI can generate content — it's that one system can do it all. Models like GPT, Gemini, and

Claude Opus are increasingly **cross-modal** — meaning they can process and relate different types of inputs and outputs. For example:

- Upload an image and get a caption or analysis
- Describe a scene and generate an illustration
- Input code and receive a narrated explanation
- Mix sound and visuals to prototype an ad

This opens the door to **true co-creation**: people expressing ideas in one form and receiving output in another — instantly, interactively, and intuitively.

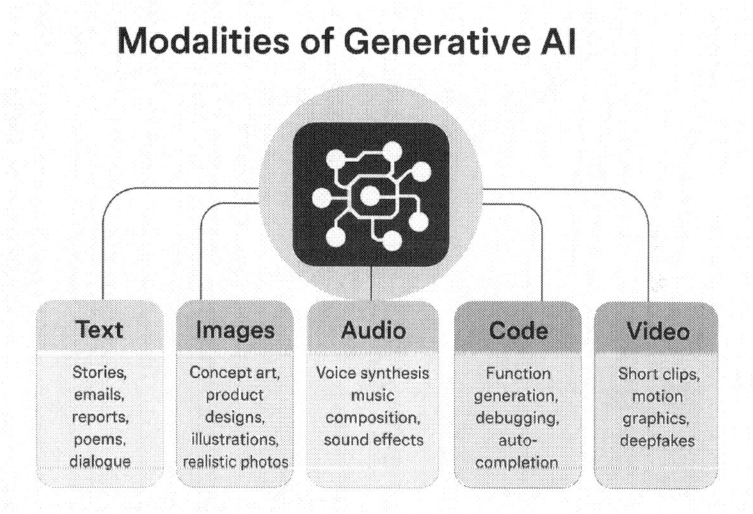

---

3. Opportunities: Creativity, Prototyping, Automation

Generative AI is already reshaping creative and technical fields in profound ways. Instead of starting from a blank page, a blank canvas, or an empty code file, people can now generate rough drafts, prototypes, or even polished pieces in seconds. The result isn't just speed — it's an expansion of what individuals and teams can imagine and produce.

Generative AI enables:

- **Rapid prototyping**: instantly generating drafts, mockups, variations, or visual concepts
- **Accessibility**: helping non-experts produce high-quality content
- **Automation**: offloading repetitive writing, coding, or design work
- **Augmentation**: supporting human creativity rather than replacing it

Across industries, these capabilities are changing workflows:

Marketing and Advertising

Brands like **Heinz** and **Coca-Cola** have used generative tools to fuel campaigns — producing thousands of AI-generated visuals from user prompts, accelerating creative exploration and sparking viral engagement.

Education and Personalized Learning

Platforms such as **Khan Academy** have integrated AI tutors that explain concepts step-by-step, adapt to a student's learning style, and provide individualized support at scale.

Entertainment, Film, and Game Development

Tools like **Runway** or **Scenario** help creators prototype scenes, characters, environments, or animations early in the production pipeline — dramatically reducing iteration time.

Design, Architecture, Product Development

Generative tools help designers brainstorm earlier and faster: producing layout options, material suggestions, style variations, 3D prototypes, and conceptual renderings in seconds.

Research and Ideation

Researchers use generative models to explore hypotheses, simulate scenarios, summarize scientific papers, or visualize datasets — augmenting the early stages of inquiry.

When used well, *generative AI becomes a collaborator,* not a replacement. It expands the range of what people can create, helps them test ideas more quickly, and frees time for the work that requires human judgment, taste, and originality.

---

4. Risks: Hallucination, Deepfakes, Misuse

Generative AI unlocks extraordinary creative and practical possibilities — but it also introduces equally extraordinary risks. Unlike traditional AI systems that merely classify or predict, generative models **produce original content**, which means they can also generate convincing misinformation, biased outputs, or harmful material. Their fluency and confidence make them powerful, but also uniquely vulnerable if used carelessly or maliciously.

Hallucinations: When AI Sounds Right But Isn't

Hallucinations occur when a generative model produces information that is false, fabricated, or logically impossible, yet presents it with complete confidence. Because these systems rely on statistical patterns rather than verified truth, they can produce output that feels reliable but is entirely incorrect.

Examples:

- A legal chatbot invents court cases and cites nonexistent rulings.
- An AI assistant generates dangerously inaccurate medical guidance.
- A document summarizer attributes quotes to authors who never said them.

Why hallucinations happen:

- **Lack of grounding**: Models generate from learned patterns, not real-time facts.
- **Training data quality**: If the data contains errors, the model reproduces them.
- **Prompt ambiguity**: Vague or open-ended prompts encourage creative guessing.

What reduces hallucinations:

- Retrieval-Augmented Generation (RAG)
- Structured prompts and constraints
- Post-generation fact-checking mechanisms

Hallucinations don't make generative AI useless — but they do mean humans must stay in the loop.

Helpful Resource — Better Prompts, Better Results

If you want a visual demonstration of how vague vs. precise prompts change AI output, the Mastering AI YouTube channel has a short tutorial that walks through real examples step-by-step.

Deepfakes: Visual and Audio Manipulation at Scale

Deepfakes use GANs, diffusion models, or similar generative methods to synthesize hyper-realistic images, audio, or video of real people — convincingly mimicking voice, appearance, and mannerisms.

Potential harms:

- **Misinformation:** Fabricated political speeches or fake news footage
- **Harassment:** Involuntary deepfake pornography and impersonation
- **Fraud:** Scam calls mimicking a CEO's voice to authorize payments

The danger is not just the content itself — but the erosion of trust. As deepfakes improve, society risks reaching a point where seeing is no longer believing.

Mitigation tools:

- AI-based detection systems
- Watermarking and provenance metadata
- Legal restrictions on malicious deepfake creation/dissemination

Bias and Stereotypes: Echoes of Inequality

Generative models learn from human data — including all its historical biases, inequalities, and stereotypes. If the training data skews toward certain perspectives, the model reflects and often amplifies them.

Examples:

- Text models that associate certain jobs with specific genders.
- Image generators underrepresenting or sexualizing certain ethnicities.
- Chatbots reproducing toxic language when prompted in harmful directions.

Bias isn't always obvious — sometimes it appears subtly in tone, examples, or representation. But its impact can be significant, especially in areas like hiring, education, or media representation.

Countermeasures:

- Diverse, representative datasets
- Bias auditing tools (e.g., IBM's AI Fairness 360)
- Human oversight in sensitive applications

Intellectual Property and Plagiarism

Many generative AI models are trained on large internet-scale datasets that likely include copyrighted material, which raises complex legal and ethical questions. This raises complex legal and ethical questions around originality, ownership, and fair use.

Examples:

- AI artwork that closely imitates a specific living artist's style
- Code generators including open-source snippets without proper attribution
- Writing tools producing passages nearly identical to published books

Key debates:

- Is AI-generated output derivative work?
- Should creators be compensated when their data trains a model?
- Who owns the output: the user, the developer, or neither?

As courts and governments catch up, organizations must establish clear internal policies for attribution, rights, and commercial use.

Malicious Use and Security Risks

Generative AI can lower the barrier to creating harmful content — empowering bad actors with tools that were once the domain of experts.

Threat scenarios:

- **Phishing campaigns** enhanced with personalized AI-generated emails.
- **Fake customer service bots** used to extract personal information.
- **Malware creation** using AI to write and obfuscate malicious code.

Organizational safeguards:

- Usage monitoring and abuse detection.
- Restrictions on fine-tuning or deploying large models in open environments.
- Internal and external red-teaming to probe AI misuse potential.

Wrapping Up: A Powerful Tool Demands Responsible Hands

Generative AI's biggest strength — its ability to **create** — is also its biggest risk. Unlike traditional software, it learns from data, mimics human output, and adjusts to user input. That means responsibility isn't only with developers; it extends to anyone who uses, deploys, or shares generative content.

Mitigating risk doesn't slow innovation.

It **enables** innovation — safely, transparently, and ethically.

---

5. From Foundation to Differentiation: Creating Value with Generative AI

Generative AI isn't just a technological advancement — it's becoming a new strategic capability. While foundation models enable powerful content generation across text, images, audio, code, and more, **the real competitive advantage comes not from using them as-is, but from how organizations adapt and apply them.**

This is the shift from *generic utility* to *differentiated value.*

Most organizations engage with generative AI in one of three ways:

1. Embedded AI Tools (Off-the-Shelf Productivity)

These are the built-in AI features inside everyday software — writing assistants, design tools, note-takers, code helpers, and more. They significantly boost productivity, but:

- They are available to everyone
- They offer no defensible advantage
- Capabilities tend to level the playing field

Useful, yes — but not strategic.

2. API Access (Building on External Intelligence)

Developers create applications that call external AI services for tasks like summarizing documents, generating images, answering questions, or producing code.

This adds flexibility, but:

- The intelligence lives outside the company
- The training data and improvements stay with the provider
- Differentiation depends more on UX than on true model behavior

APIs unlock creativity — but the **value** remains with the model owner.

3. AI Platforms (The Path to Real Differentiation)

This is where generative AI becomes transformative. Organizations start with a pretrained foundation model, then **customize it with their own data, tasks, terminology, and workflows.**

This approach turns a general-purpose model into a unique creative engine that reflects the company's expertise and identity.

This can include:

- **Retrieval-Augmented Generation (RAG)** to ground outputs in real-time or domain-specific information
- **Fine-tuning** on internal documents like manuals, reports, call transcripts, or case histories
- **Adapters or LoRA techniques** to add new capabilities without touching the base model
- **Internal copilots** customized for teams (marketing, legal, HR, product)
- **Customer-facing tools** built on tailored generative capabilities (chatbots, writing assistants, design engines)

When done well, this shifts AI from "a tool everyone uses" to **a proprietary asset only you have.**

This is where value is created — not from the foundation model itself, but from how it is shaped, governed, and deployed.

Ownership, Governance, and Trust

As organizations build these systems, strong governance becomes essential:

- **Transparency** about what data was used
- **Clear oversight** of behavior and outputs
- **Defined guardrails** for accuracy, safety, and tone
- **Alignment with brand values and responsible use guidelines**

Strategic AI is not just about what the model can do — it's about how it behaves, how it is controlled, and how reliably it supports the organization's real-world processes.

The Future: Many Models, Many Creators

The future of generative AI will not be dominated by a single model or a single approach. Instead, it will be shaped by:

- Many models
- Many datasets
- Many workflows
- Many layers of customization

The organizations that win will be the ones that don't **settle for generic outputs.**

They will build systems tailored to their voice, their knowledge, and their mission.

Generative AI's true power lies not in what it can produce automatically — but in the unique value organizations unlock when they make it their own.

---

6. Wrapping Up: Generation Is Not Understanding

Generative AI has ushered in a new era. For the first time, we have systems that don't just recognize patterns — they invent. They write essays, compose music, produce images, draft code, and generate videos. This creative capability marks a profound shift from earlier forms of AI.

But as impressive and human-like as these outputs can seem, it's essential to remember a core truth:

Generative models do not understand what they create.

They operate by predicting what is likely to come next based on patterns in data — not by reasoning, comprehension, or intent. This distinction matters, because it explains both their strengths and their limitations.

Where generative AI excels

Generative models are unmatched in:

- speed
- scale
- fluency
- flexibility
- variation and creativity

They can produce drafts, ideas, prototypes, and explanations far faster than any human — and do it consistently.

Where it falls short

But these strengths come with weaknesses:

- They can produce factually wrong information with confidence.
- They can replicate bias or unfair assumptions.
- They can generate content that appears thoughtful but lacks real understanding.
- They struggle with nuance, ethics, intent, and context unless guided carefully.

This is why **the best outcomes come from human–AI collaboration**.

AI can generate possibilities, but humans provide judgment: deciding what's accurate, what's ethical, what's useful, and what aligns with our values.

Generative AI is not just another feature. It's a new frontier — a technological shift comparable to the arrival of the internet or the smartphone. Those who treat generative systems as strategic tools, rather than novelties, will shape the next wave of creative, educational, and industrial innovation.

The promise is enormous — but it requires clear understanding and responsible use.

Quick Takeaways: Generative AI

» Creates, not just classifies

Generative AI models go beyond recognizing or labeling — they produce original content like text, images, and code.

» Powered by modern architectures

Core technologies include **Transformers**, **Diffusion Models**, and **GANs** — each enabling different kinds of generative outputs.

» **Multimodal capabilities**

These models can generate across modalities: **language, visuals, audio, code, and even video**, often from the same core architecture.

» **Enhances creativity and productivity**

From brainstorming and prototyping to automating repetitive work, generative AI supports faster, more imaginative workflows.

» **Customization creates real value**

True differentiation comes from tailoring foundation models with your **own data, tools, and workflows** — turning AI into a strategic asset.

» **Risks must be managed**

Hallucinations, deepfakes, bias, and misuse are serious concerns. **Governance and oversight** are essential for responsible use.

» **Pattern, not understanding**

Generative AI mimics human output but doesn't "understand" meaning — reminding us to use it thoughtfully and critically.

Chapter Summary

This chapter explored the major shift from AI systems that recognize patterns to systems that generate entirely new content. We examined the core architectures that enable this transformation — from transformers to diffusion models — and how they allow generative AI to produce text, images, audio, code, and even early-stage video.

We then looked at the practical impact of these capabilities: faster prototyping, expanded creativity, automation of repetitive tasks, and new forms of human–AI collaboration. These benefits extend across industries, from design and marketing to education, entertainment, and software development.

But generative AI also introduces meaningful risks: hallucinations, misinformation, deepfakes, embedded bias, intellectual property concerns, and malicious uses. As these tools become more accessible, responsible deployment, governance, and human oversight become essential.

A key theme of the chapter is differentiation. True value emerges not from using foundation models "as-is," but from adapting them — through fine-tuning, RAG, custom workflows, proprietary data, and internal AI platforms. Organizations that invest in building their own generative capabilities will unlock competitive, creative, and operational advantages.

Ultimately, generative AI does not "understand." It predicts. Its power lies in speed, scale, and fluency — and its impact depends on how humans guide, verify, and apply what it produces. Used thoughtfully, it expands what's possible in creativity, communication, and innovation.

Chapter 12

HOW AI UNDERSTANDS THE WORLD

Introduction: From Input to Insight

Most of us interact with AI through outputs: a chatbot response, a recommended image, a generated summary. But behind every output lies a critical first step — **perception**. Before an AI system can create, decide, or assist, it must first understand what it's being given: to read, to listen, to see.

This chapter explores how artificial intelligence processes and interprets different types of input — text, images, sound, and even multiple modalities at once. These capabilities form the bridge between raw data and meaningful action. This is where AI shifts from simply reacting to the world to interpreting it.

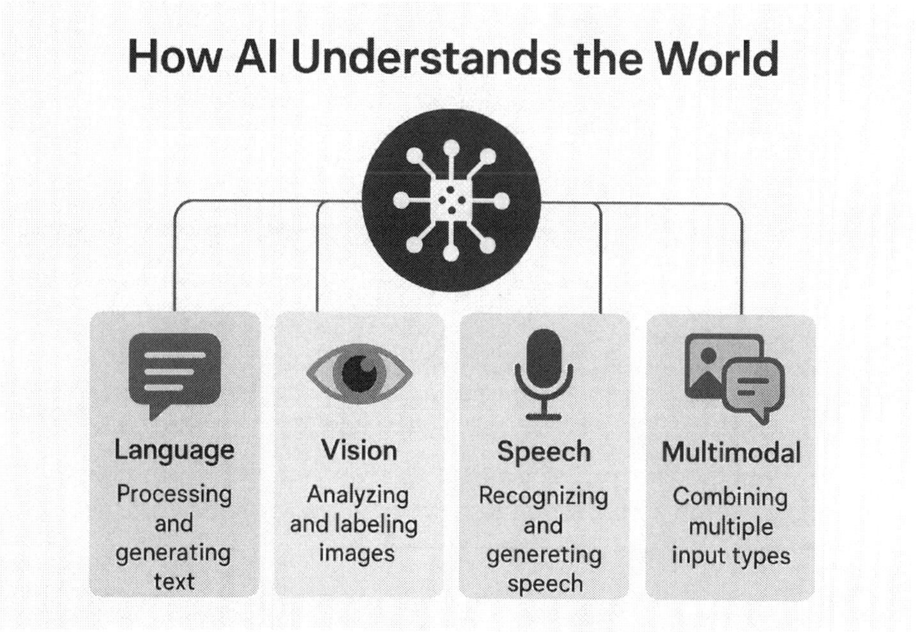

1. Language (NLP): How Models Read and Write

Natural Language Processing (NLP) is the branch of AI that enables computers to understand, interpret, and generate human language. You use it every time you interact with a chatbot, request a translation, or ask a voice assistant for help. But how do machines make sense of something as complex, ambiguous, and context-dependent as human language?

At its core, NLP bridges **unstructured text** — the natural, flowing way we speak and write — with the **structured representations** machines need to process and act on. For example, a command like *"Add eggs and milk to my shopping list"* is perfectly clear to a human, but a computer needs to turn that into a structured list with the items labeled "eggs" and "milk."

This translation process happens in two directions:

- **Natural Language Understanding (NLU)**: Converting human language into structured data a computer can use.
- **Natural Language Generation (NLG):** Taking structured data and turning it back into natural-sounding language.

While many applications use both directions, this section focuses on how NLP systems understand us — the NLU side.

Core Tools of NLP

NLP isn't powered by a single algorithm. It's a layered toolkit of methods working together to extract meaning from text:

- **Tokenization:** Splits text into individual words, subwords, or phrases ("tokens"), allowing the system to analyze language piece by piece.
- **Stemming and Lemmatization:** Reduce words to their base form. For example, "runs," "ran," and "running" may all map to "run." Lemmatization goes further, using grammar rules and dictionaries to map "better" to "good."
- **Part-of-Speech Tagging:** Identifies whether a word is a noun, verb, adjective, etc., based on its role in the sentence — since words like "make" can function differently depending on context.
- **Named Entity Recognition (NER):** Detects proper nouns like people, places, or organizations (e.g., recognizing "Arizona" as a state or "Ralph" as a person).

Together, these tools move machines from *raw text* toward *structured* meaning.

Real-World Applications

Natural Language Processing powers a broad range of everyday technologies:

- **Chatbots and virtual assistants**, which translate spoken or typed words into actions.
- **Machine translation**, where nuance, tone, and context matter — avoiding awkward mistranslations.
- **Sentiment analysis**, used in marketing and customer service to detect tone, satisfaction, or frustration.
- **Spam detection**, where models identify patterns like urgency, suspicious phrasing, or poor grammar.

Powered by Transformers

Modern NLP is driven by transformer architectures — the same kind behind large language models (LLMs) such as ChatGPT, Gemini, Claude, and others.

Unlike earlier approaches that read text word-by-word, transformers analyze entire sequences at once and model how words relate to one another in context. This allows them to capture subtle patterns in grammar, meaning, and tone.

These models are trained on enormous datasets — books, websites, articles, reviews — learning statistical patterns of how humans write and speak. The result is systems that can generate fluent, often strikingly coherent text by predicting the most likely next word based on everything that came before.

Key Concept

Language models do not "understand" meaning in the human sense. They don't know what words represent — they recognize how words tend to appear together, and use those patterns to generate plausible responses.

Despite this limitation, the results can be remarkably useful and context-aware.

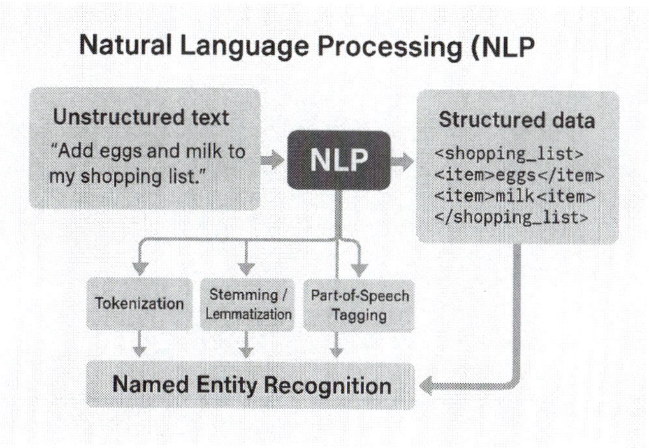

2. Vision: How AI Sees the World

Computer vision is the branch of artificial intelligence that enables machines to interpret and understand visual information — images, videos, and real-world scenes. But while humans see with context, intuition, and lived experience, machines operate by analyzing **patterns, pixels, and geometry.**

At its core, computer vision aims to derive meaning from visual data. This includes identifying objects, classifying images, detecting anomalies, and even generating new visuals. The field has evolved from early rule-based edge detection to powerful neural architectures that learn patterns automatically.

Today's most effective systems are driven by two key architectures:

- **Convolutional Neural Networks (CNNs)** — Specialized networks tuned for image data. Each layer detects increasingly abstract features: edges → textures → shapes → full objects.
- **Vision Transformers (ViTs)** — A newer architecture that applies attention mechanisms to visual tasks. Instead of scanning an image step-by-step, ViTs divide images into patches and analyze them holistically, capturing spatial relationships and context more effectively.

Computer vision is not just about what is in the image — but also how visual elements relate to one another, and how a system can act based on what it recognizes.

Applications of Computer Vision

Computer vision is deeply embedded across industries, enabling automation and insights at scale:

- **Facial recognition:** For phone unlocking, secure authentication, or identity verification.
- **Autonomous vehicles:** Interprets road signs, detects pedestrians, and maps surroundings to support safe navigation.
- **Medical diagnostics:** Analyzes X-rays, MRIs, and CT scans to assist doctors in detecting diseases.

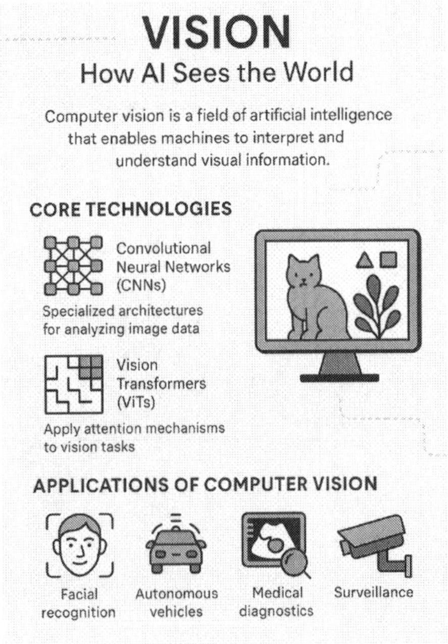

- **Retail and logistics:** Supports automated checkout, inventory tracking, and product recognition.
- **Manufacturing:** Performs real-time defect detection and quality control.
- **Agriculture:** Monitors crop health through drone imagery and automates field analysis.
- **Security and surveillance:** Tracks movement and identifies unusual patterns across camera networks.

These use cases highlight both the breadth and precision of modern computer vision.

Techniques Within Computer Vision

Several core tasks make up the foundation of computer vision systems:

- **Image Classification:** Assigns a label to an entire image (e.g., "cat," "dog," "tumor").
- **Object Detection:** Identifies and locates specific objects in an image (e.g., detecting all cars on a street) using models like YOLO or Faster R-CNN.
- **Image Segmentation:** Breaks an image into pixel-level regions, offering more detailed understanding — essential for medical imaging or self-driving perception.
- **Facial Analysis:** Detects faces and maps facial landmarks to analyze identity, expression, or emotion.

Each technique adds a layer of understanding, moving from broad categorization to fine-level interpretation.

Strengths and Limitations

Computer vision excels at pattern recognition at scale — spotting details humans might miss or processing thousands of images with consistent accuracy. But it has notable limits.

- A vision model can identify a dog in a photo — but it cannot infer why the dog looks frightened or that it's reacting to something outside the frame.
- It does not "see" in the human sense — it interprets mathematical representations of pixels.
- Context and reasoning remain difficult unless paired with language models in multimodal systems.

As architectures like vision transformers, contrastive models (e.g., CLIP), and multimodal training evolve, AI systems are becoming more capable of aligning images with meaning. Yet their performance still depends heavily on training data, diversity, and careful design — not intuition or experience.

3. Speech: How AI Listens and Talks

Spoken language is one of the most natural forms of human communication. But to a computer, sound is nothing more than vibration — raw, unstructured, and unintelligible. Speech technology bridges that gap by converting **spoken words into text** and then converting **text back into natural-sounding speech**.

This two-way loop is what makes voice-enabled systems feel conversational. When you ask a smart speaker for the weather or dictate an email, you're relying on the interaction between **Speech-to-Text (STT)** and **Text-to-Speech (TTS)** — both powered today by deep learning.

Speech-to-Text (STT): From Voice to Text

STT models transcribe spoken language into written text. They're trained on enormous datasets of audio paired with human transcripts, learning to map sound waves to phonemes, words, and full sentences.

Modern STT systems perform several tasks in milliseconds:

- Break speech into phonetic units
- Detect word boundaries
- Handle accents, pacing, slang, and intonation
- Filter out background noise
- Adapt dynamically to different speakers

Applications of STT include:

- Live captioning (e.g., YouTube subtitles, accessibility tools)
- Voice assistants (Siri, Google Assistant, Alexa)
- Voice search ("Play jazz," "Turn off the lights")
- Call center transcription and analytics
- Dictation tools and meeting transcription services

The improvements in STT accuracy over the past decade — especially for multilingual and accented speech — are almost entirely due to advances in deep learning architectures.

Text-to-Speech (TTS): From Text to Voice

TTS systems perform the reverse transformation: they take structured text and produce synthetic speech. Early systems sounded robotic, but modern neural TTS can clone voices, mimic emotion, and adjust tone or style.

Most TTS systems combine two components:

1. **A voice model** — learns a speaker's timbre, pacing, and pronunciation from a set of recordings.
2. **A synthesis model** — generates audio from text and refines it to match the target voice.

Applications include:

- Virtual assistants and smart devices
- GPS and navigation systems
- Audiobook and voiceover production
- Accessibility tools for users with speech limitations
- Customer service agents and automated hotlines

Many high-end TTS models can produce speech that, in short clips and everyday use, is difficult for most listeners to distinguish from a real human voice — raising both creative possibilities and ethical questions.

Integrating NLP, STT, and TTS

When combined with language models, STT and TTS enable true human–AI interaction:

- **STT** captures the spoken input
- **NLP** interprets intent, context, and meaning
- **TTS** produces a spoken response

This enables complex real-time interactions:

- Asking a voice assistant in Spanish → getting a response in Spanish
- Speaking into a translation device → hearing the translation aloud
- Dictating an email → hearing it read back before sending

Behind the scenes, multimodal models coordinate this workflow seamlessly — converting speech to text, text to meaning, meaning to text, and text back to speech within seconds.

Strengths and Challenges

Speech systems have become impressively accurate and expressive, but they face significant challenges:

1. Linguistic diversity

Accents, dialects, and regional variations can cause transcription errors. A system trained primarily on American English will struggle with Scottish, Indian, or Colombian accents unless the dataset is diverse enough.

2. Environmental noise

Crowded rooms, cars, wind, or overlapping conversations degrade STT quality — especially in real-time.

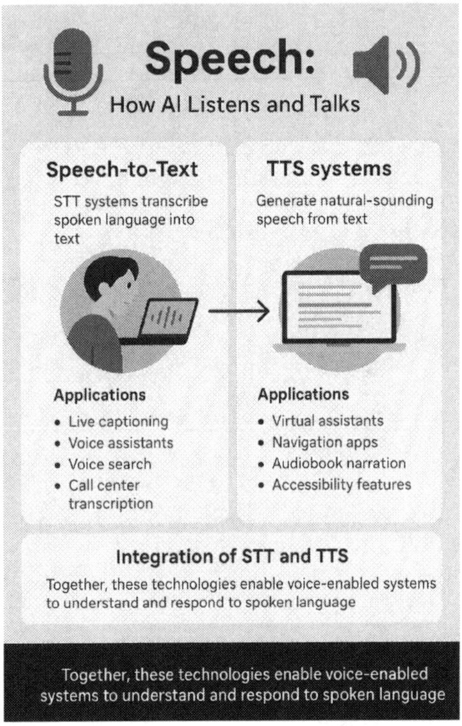

3. Emotional nuance

TTS systems are improving, but subtle emotional tones can still sound flat or uncanny unless the model is carefully trained or fine-tuned.

4. Privacy and trust

Since many devices process speech in the cloud:

- Users may not know what gets recorded
- Some systems retain voice samples
- "Always-on" microphones raise surveillance concerns

These issues make transparency, consent, and data governance essential for voice-enabled technologies.

Despite the challenges, voice interaction has rapidly evolved from novelty to default. Increasingly, users expect technology to understand them when they speak — not just when they type.

---

4. Multimodal Systems: When AI Combines Senses

Humans rarely rely on a single sense to understand the world. We look, listen, read, and interpret these inputs together. AI is beginning to do the same through multimodal systems — models that integrate text, images, audio, video, and other signals into a unified understanding.

These systems don't treat language, vision, or sound as isolated channels. Instead, they learn how modalities interact: how a caption relates to a photo, how tone can change the meaning of speech, or how a chart supports the argument of a document. This fusion allows them to handle richer, more complex tasks than any single-modality model could.

What Makes Multimodal AI Different?

Traditional AI systems are specialized:

- A vision model processes images
- A language model analyzes text
- A speech model transcribes audio

Multimodal AI blends these capabilities, creating shared internal representations across modalities. This allows the model to understand not just what is seen or written, but how those inputs relate.

Modern multimodal systems are often built on transformer architectures adapted to handle sequences of different data types. Examples include:

- GPT with Vision, Gemini, or Claude — which accept text and images together
- CLIP (Contrastive Language–Image Pre-training) — which aligns text and image meaning
- Vision–Language Action models used in robotics

By learning connections between modalities, these systems enable cross-modal reasoning that feels far more intuitive and flexible.

Capabilities of Multimodal AI

Multimodal models already enable a wide variety of real-world applications:

- **Image captioning:** Generating accurate descriptions of pictures.
- **Visual question answering:** Answering questions about an image (e.g., "How many dogs are in this photo?").
- **Chart or diagram analysis:** Interpreting visuals in documents or scientific papers.
- **Video understanding:** Summarizing, tagging, or generating commentary from video content.
- **Assistive technology:** Helping blind users "see" by narrating visual surroundings.
- **Search and retrieval:** Combining text queries with visual elements (e.g., "Show me this product in red").
- **Customer Support:** Platforms like Zendesk integrate multimodal tools to analyze screenshots, audio, and chat history for better issue resolution.

- **Conversational AI:** Replika and Character.AI combine chat logs and visual input to build emotionally responsive digital companions.
- **Accessibility:** Apps like Be My Eyes use GPT Vision to describe the environment to blind users in real time.

These use cases require a blend of linguistic reasoning, visual understanding, and contextual alignment — exactly what multimodal systems are designed to deliver.

The Rise of Multimodal Chatbots

Generative AI tools like ChatGPT, Gemini, and Claude are rapidly evolving into multimodal assistants. They can now:

- Accept image uploads and analyze the content
- Read and interpret charts, forms, and screenshots
- Extract information from scanned documents
- Generate images or diagrams from text
- Describe visuals in natural language
- Combine audio, text, and image reasoning in a single conversation

This makes AI useful in a much wider range of tasks, from diagnosing technical problems (via screenshots) to interpreting contracts (via scanned pages) to brainstorming design ideas (via sketches).

Limits and Challenges

Despite rapid progress, multimodal AI still faces meaningful constraints:

Context fragmentation

Combining inputs doesn't ensure full comprehension. A model may accurately describe objects in an image but misunderstand their significance.

Ambiguity

Visuals often lack labels or explicit context. A photo of a person looking surprised might reflect joy, fear, or shock — distinctions that require human judgment.

Computational demands

Multimodal systems require enormous compute resources and highly curated datasets, making them expensive to train and deploy.

Bias across modalities

Combining text + images can amplify stereotypes in both directions. For example, biased textual associations paired with skewed image datasets can reinforce unintended patterns.

Even with these challenges, multimodal AI represents a major step toward systems that perceive the world in ways closer to how humans naturally do — integrating language, vision, sound, and context into a more coherent whole.

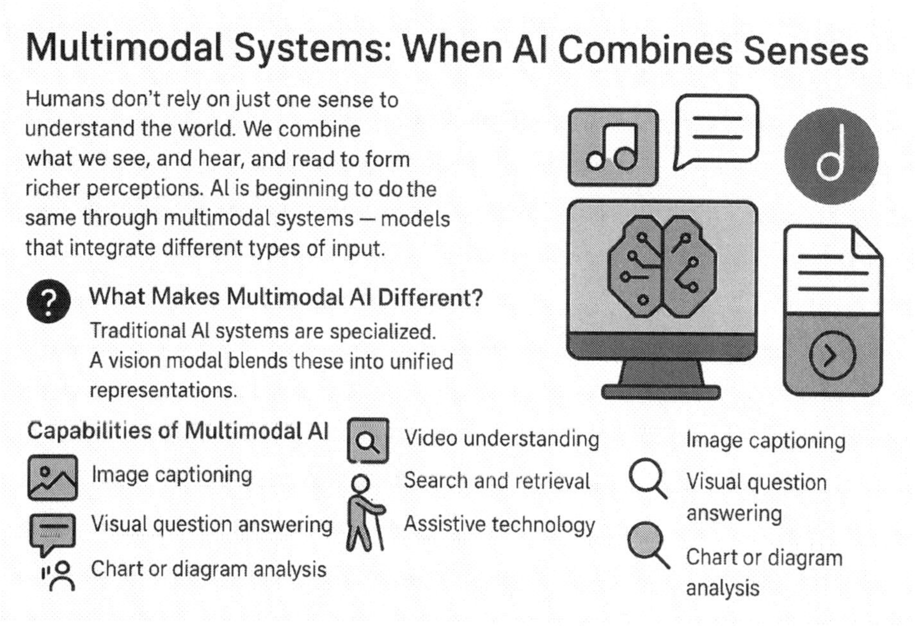

5. Sensation vs. Interpretation: The Limits of Understanding

AI can now *perceive* the world in increasingly sophisticated ways — reading text, recognizing images, transcribing speech, and even combining modalities to form richer inputs. But perception is not understanding. Models don't experience meaning the way humans do.

A system might identify a face, summarize a paragraph, or repeat a melody, but it doesn't know what those things signify. It processes patterns — not experiences.

This distinction matters.

Humans interpret the world through lived experience, emotional context, intuition, cultural knowledge, and common sense. AI lacks all of these. It does not feel surprise when it hears a sudden noise, or empathy when it sees someone crying. It can label "smile," but it doesn't understand why someone is smiling — or whether that smile is joyful, nervous, or forced.

Why the distinction is critical:

- **Overtrust.** People may assume the AI "understands" because its outputs appear fluent or insightful.
- **Misuse.** Without recognizing AI's limits, users may rely on systems for decisions that require human judgment or emotional nuance.
- **Ethical blind spots.** Misinterpreting AI's abilities can lead to flawed deployments — especially in high-stakes areas like healthcare, hiring, law enforcement, or education.

Even as multimodal models become more capable — seeing, reading, listening, and describing — they still lack the human faculties required for interpretation. They do not infer motives. They do not form beliefs. They do not understand consequences. They act on statistical associations, not comprehension.

As AI perception improves, staying mindful of this boundary between pattern recognition and meaning becomes increasingly important. Recognizing these limits helps us deploy AI responsibly — using it as a powerful tool, not as a substitute for human insight.

Quick Takeaways: How AI Understands the World

» Language (NLP)

NLP enables AI to process and generate human language — not with true comprehension, but through learned linguistic patterns.

» Vision

Computer vision systems analyze images and video, supporting applications from facial recognition to medical diagnostics.

» Speech

Speech technologies convert spoken language to text and vice versa, enabling voice interaction and audio synthesis.

» Multimodal Systems

These models integrate multiple input types — like text, images, and audio — to support richer, more context-aware tasks.

» Limits of Perception

AI perceives patterns, not meaning. It lacks emotional context, lived experience, and the true understanding behind human communication.

Chapter Summary

This chapter explored how modern AI systems perceive the world through four major capabilities: natural language processing, computer vision, speech recognition and synthesis, and multimodal integration. These perception abilities form the foundation for many of today's generative and decision-making systems — enabling them to read text, interpret images, understand speech, and combine multiple modalities into coherent responses.

We saw how **NLP** turns raw text into structured data, how **computer vision** extracts meaning from images and videos, how **speech technologies** bridge the gap between spoken and written language, and how **multimodal models** unify different forms of input to support richer interactions. Across all these domains, transformers and deep learning architectures enable more sophisticated pattern analysis than ever before.

Yet, perception is not understanding. AI systems don't interpret the world through intuition, emotion, or shared human experience. They detect patterns — powerful, useful patterns — but without awareness or judgment. Recognizing the boundary between what AI can perceive and what it can truly understand is essential for using these systems responsibly and effectively.

Chapter 13

HOW AI LEARNS OVER TIME

Introduction: Teaching the Machine — Then and Now

Artificial intelligence isn't static. It begins with training, but its real value comes from how well it can adapt, specialize, and improve. Just as human learning continues long after school, AI learning doesn't end once the initial training phase is complete.

In earlier chapters, we explored how models are pretrained on massive datasets and then applied to various tasks. But that first phase is only the foundation. In real-world environments, AI systems often need to learn new things, adapt to changing conditions, or update themselves based on feedback. If they don't, they risk becoming outdated, biased, or ineffective.

This chapter explores how AI evolves after its initial training — through techniques like fine-tuning, reinforcement learning, transfer learning, and continual learning. We'll also look at how AI learns from small numbers of examples, how it chooses what to learn next, and how it can forget or drift away from its original accuracy. These concepts reveal how modern AI becomes more than a one-time model — it becomes a system that grows, adapts, and responds to the world around it.

---

1. Pretraining, Fine-Tuning, and Reinforcement Learning

Modern AI often starts with **pretraining** — a process in which a model is trained on massive amounts of general-purpose data, often in an unsupervised or self-supervised way. This gives the model a broad understanding of language, images, or code before it's ever applied to a specific task.

Pretraining

Think of pretraining like sending a student to read the entire internet before giving them a job. The model absorbs billions of words, images, or other inputs to develop a sense of how patterns work—how sentences are formed, how images are structured, how code is written. It doesn't receive specific guidance; it simply learns to predict what comes next.

For example:

- GPT-style language models are pretrained to predict the next word in a sentence.
- Vision models are trained to associate patterns of pixels with labels or categories.

Pretraining is extremely compute-intensive and expensive — but it lays the foundation for everything that follows.

Fine-Tuning

After pretraining, the model can be **fine-tuned** — trained further on a smaller, domain-specific dataset. This is like taking a well-read student and teaching them legal writing, medical terminology, or customer service skills.

Fine-tuning helps the model **specialize**, adjusting its broad knowledge to fit specific needs.

Example:

A general model can be fine-tuned on thousands of legal contracts so it becomes better at summarizing, reviewing, or answering questions about them. This approach is far faster and more efficient than training a new model from scratch.

Reinforcement Learning

In contrast to pretraining and fine-tuning — which use fixed datasets — **reinforcement learning (RL)** involves learning through trial and error. The model takes actions, receives feedback in the form of a reward signal, and updates its behavior over time.

Reinforcement learning is used in:

- **Game-playing agents** (e.g., AlphaGo, which learned by playing millions of games against itself)
- **Robotics**, where agents learn by physically interacting with their environment
- **Dialogue systems**, such as ChatGPT, whose behavior is refined using Reinforcement Learning from Human Feedback (RLHF)

Not all AI systems use reinforcement learning, but it's essential for tasks where decisions lead to long-term outcomes or delayed rewards.

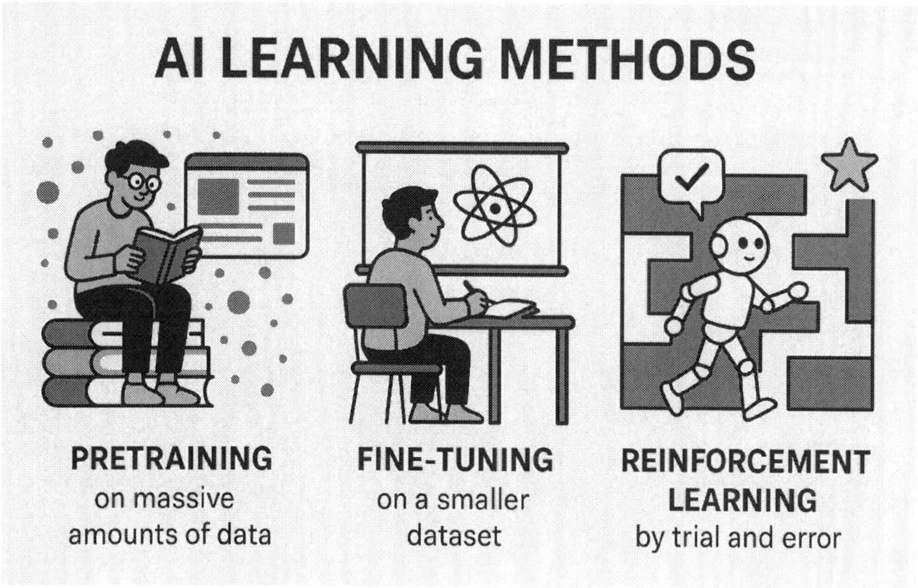

2. Transfer Learning and Few-Shot Learning

Once a model has learned something useful, can it apply that knowledge elsewhere? This is the core idea behind transfer learning — one of the most important developments in modern AI. Instead of starting from scratch every time, we reuse what a model has already learned and adapt it to a new task.

Transfer Learning: Reusing Knowledge

In traditional machine learning, models were trained in isolation — one model per task, one dataset at a time. But deep learning models are often **overparameterized**, meaning they learn general features (edges, shapes, sentence structures, semantic patterns) that can apply across many different contexts.

Transfer learning allows a pretrained model to become a foundation for a related task. Examples include:

- A vision model trained to identify animals adapted to identify medical scans.
- A language model trained on books and websites adapted to summarize legal documents or answer financial questions.

This makes model development **faster, cheaper, and more accessible**, because teams no longer need massive labeled datasets or enormous compute to build a strong model. They start with a foundation and adapt it.

Few-Shot and Zero-Shot Learning

Thanks to transfer learning (and the scale of pretrained models), modern systems can perform surprisingly well with minimal examples. This is where **few-shot** and **zero-shot** learning come into play.

Few-shot learning

The model is given just a handful of examples, then asked to generalize.

Zero-shot learning

The model receives no examples — just a clear instruction — and relies on its pretrained knowledge to perform the task.

For example, you could simply say:

"Translate this sentence to French."

The model does so — even if it hasn't been explicitly trained with that exact prompt structure — because it has learned broad linguistic patterns during pretraining.

Few-shot learning has become a cornerstone of prompt engineering, the practice of designing prompts that guide AI behavior. It allows people to prototype tools, automations, and workflows quickly without retraining or fine-tuning the underlying model.

Why This Matters

Few-shot and zero-shot learning dramatically lower the barrier to experimentation. You don't need a massive labeled dataset or specialized technical knowledge to make AI useful. This flexibility — powered by transfer learning — is why models like GPT, Claude, and Gemini can adapt so easily to a wide range of tasks without custom code.

It is one of the reasons generative AI exploded so quickly: the same foundation can power thousands of applications, each shaped by instructions, examples, or small amounts of domain-specific data.

---

3. Continual Learning — Can AI Keep Learning?

One of the defining traits of human intelligence is adaptability. We don't need to start from zero each time we encounter a new task — we build on past experience and improve as we go.

So can AI do the same?

The answer is... **partially.** The concept of continual learning (also called lifelong learning) explores whether AI systems can learn incrementally, without forgetting what they already know.

The Challenge of "Catastrophic Forgetting"

Most traditional AI models are static: once trained, they don't learn new things unless they are retrained from scratch. When you try to update a model with new data, it often suffers from catastrophic forgetting — overwriting previous knowledge in favor of the new patterns.

This is a major challenge for AI that needs to adapt continuously, such as:

- Chatbots that learn from new customer interactions
- Fraud detectors adapting to new attack patterns
- Personalized assistants that remember user preferences over time

Without strategies to protect old knowledge, updating the model can degrade earlier skills or accuracy.

Strategies for Continual Learning

Researchers have developed several approaches to help models retain old skills while learning new ones:

Replay-based methods

Store a subset of the model's past training data and mix it with new data during updates. This gives the model reminders of what it previously learned.

Regularization methods

Penalize the model for changing certain weights too drastically — especially those known to be important for prior tasks.

Dynamic architecture methods

Expand the model's capacity over time, allowing it to learn new concepts without overwriting older representations.

While these approaches show promise, they are still primarily research-driven and not yet widely used in large commercial systems. Today, most production AI relies on a simpler approach: periodic batch retraining on updated datasets.

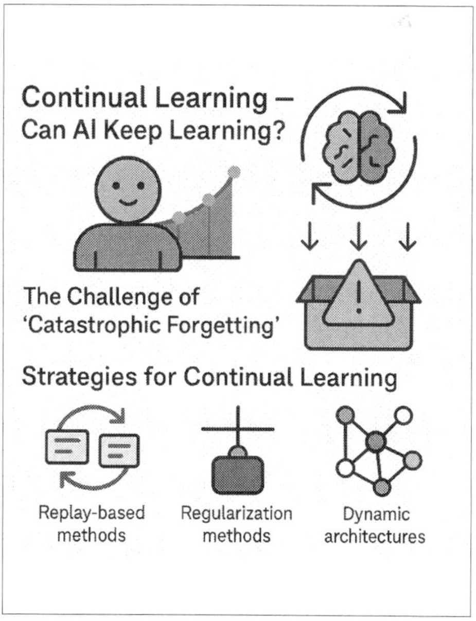

Real-World Implications

Continual learning is essential for building adaptive, interactive AI systems. Imagine:

- A customer support agent that gets better with every conversation
- A recommendation system that evolves as your tastes change
- A robot that continually learns about its environment and improves its behavior

These examples require models that evolve continuously — not static snapshots.

But this flexibility also introduces new challenges:

- How do we store or summarize what the model has learned?
- How do we prevent drift or degradation over time?
- How do we ensure updates don't introduce bias or instability?

The long-term goal is an AI that learns steadily and responsibly, more like a human — flexible, contextual, and capable of growing without losing its foundations. We're not there yet, but the frameworks being developed today form the basis for that future.

---

4. Active Learning and the Role of Uncertainty

One of the biggest bottlenecks in machine learning is labeled data. While large foundation models benefit from vast unsupervised datasets, many real-world applications — from medical diagnosis to document classification — still depend on high-quality human labels. But labeling everything is expensive, time-consuming, and often unnecessary.

Active learning offers a more efficient path: instead of learning from all available data, the model selectively chooses which examples to label — specifically the ones it is *most uncertain* about.

What Is Active Learning?

Active learning is a training strategy in which the AI model helps determine which data points would be most valuable to label next.

Rather than training blindly on a random dataset, the model essentially asks:

"Which examples confuse me the most? Show me the answers to those."

The loop works like this:

1. The model is trained on a small labeled dataset.
2. It scans a large pool of unlabeled data and flags examples where its predictions are least confident.
3. A human labels those uncertain cases.
4. The new labeled data is added back into the training set, and the model is updated.
5. The cycle repeats.

This lets the model learn quickly from the right data, not just more data — especially powerful in fields where labels require time, money, or expert review.

Measuring Uncertainty

But how does a model know it's uncertain?

Several indicators help:

Low confidence scores

If the model assigns similarly low probabilities to all labels, it's signaling confusion.

Disagreement among models

Ensembles or model variations disagreeing on the same input often point to ambiguity.

Entropy

A statistical measure of unpredictability — higher entropy means greater uncertainty in predictions.

These signals help the model identify where human labeling would have the biggest impact.

Where Active Learning Shines

Active learning is especially useful when expert judgment is required or when edge cases matter:

- **Healthcare:** Flagging ambiguous X-rays, MRIs, or pathology slides for radiologist review.
- **Legal Technology:** Prioritizing unclear clauses or contracts for attorney annotation.
- **Customer Support:** Escalating uncertain chatbot queries to human agents and learning from those examples.
- **Autonomous Vehicles:** Identifying rare, unusual, or dangerous traffic scenarios that need manual annotation.

In all of these settings, active learning reduces labeling burden while concentrating effort where it counts most.

Why It Matters

Active learning mirrors how humans learn: we focus on gaps in our understanding, not on what we already know.

By embracing uncertainty rather than ignoring it, AI can:

- Learn more efficiently
- Adapt more quickly
- Reduce reliance on massive labeled datasets
- Improve accuracy with fewer resources

It's not about feeding a model more data — it's about feeding it the right data at the right time.

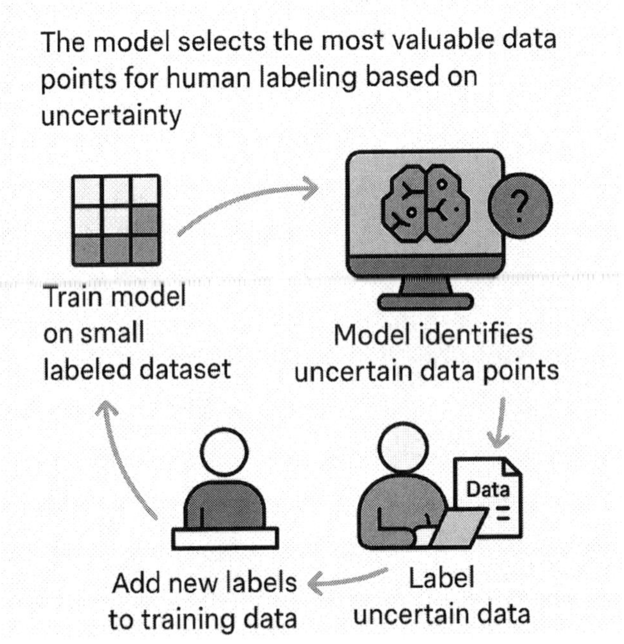

5. Model Drift and Lifecycles

Training an AI model isn't a one-time event — it's the start of an ongoing cycle. As the world changes, so does the data flowing through AI systems. When real-world inputs shift away from the patterns the model was trained on, performance can decline. This phenomenon is known as **model drift**, and it's one of the most important — and often underestimated — challenges in AI deployment.

What Is Model Drift?

Model drift occurs when an AI system's predictions become less accurate because the data it encounters over time begins to differ from the data it originally learned from. There are two main forms:

Data drift (covariate shift):

The input distribution changes.

Example: a spam filter trained on 2021 email tactics performs poorly against new phishing strategies in 2025.

Concept drift:

The relationship between inputs and outputs changes.

Example: in fraud detection, transaction patterns once seen as suspicious may later reflect legitimate behavior as customer habits evolve.

Unchecked drift can lead to poor decisions, safety risks, or loss of trust — especially in high-stakes areas like finance, healthcare, or cybersecurity.

The AI Lifecycle

Because of drift, organizations must treat AI as a living system that requires continuous care. A healthy AI lifecycle typically includes:

Data Monitoring:

Track the characteristics of incoming data. Are patterns shifting? Are new trends emerging?

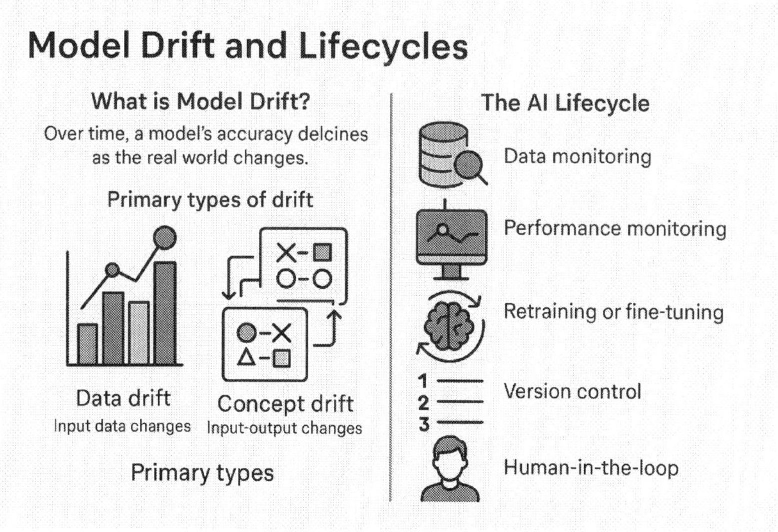

Performance Monitoring:

Measure real-world accuracy, user feedback, and error patterns.

Sudden drops often signal drift.

Retraining or Fine-Tuning:

Update the model periodically using new data — either incrementally or through full retraining. This keeps the system aligned with current conditions.

Version Control:

Maintain multiple model versions so teams can compare results, audit decisions, or roll back problematic updates.

Human-in-the-loop Oversight:

In sensitive domains, human review acts as a safeguard. Humans catch mistakes, validate decisions, and provide labeled examples for future improvements.

These processes are core to responsible, long-term AI deployment.

Industries Where Drift Matters Most

Model drift has direct consequences in many sectors:

- **Finance:** adapting risk models or trading systems to changing market behavior
- **Healthcare:** adjusting diagnostic models as diseases evolve or imaging devices change
- **Retail:** updating recommendations to reflect seasonal or cultural shifts
- **Cybersecurity:** detecting new forms of attacks, anomalies, or malware patterns

In each case, drift can affect both performance and trust.

Global Governance and Regulation

As AI systems grow more adaptive, regulatory and governance frameworks are evolving to ensure transparency, fairness, and accountability. Around the world:

Canada

Developing the Artificial Intelligence and Data Act to balance innovation with human rights.

Singapore

Implemented the Model AI Governance Framework, providing practical guidance for responsible deployment.

Brazil

Debating national legislation emphasizing transparency, non-discrimination, and oversight.

Alongside the **EU AI Act** and the **U.S. NIST framework**, these efforts reflect a global shift: as AI becomes more dynamic, legal systems must adapt just as quickly.

Why It Matters

Model drift isn't just a technical challenge — it's a trust and governance challenge.

If systems degrade silently, organizations may make poor decisions without realizing the model has become outdated.

Building responsible AI means thinking beyond the launch:

- setting up monitoring infrastructure,
- establishing retraining workflows, and
- building human oversight into the lifecycle.

AI must evolve at the pace of the world it operates in — and that requires ongoing attention, not one-time training.

---

6. Wrapping Up: Teaching the Machine, Continuously

When we think of machine learning, it's easy to imagine a single moment in time: gather data, train a model, deploy it — done. But real-world AI doesn't work that way. Learning is not a checkbox. It's a cycle.

Just as humans learn and adjust throughout their lives, AI systems improve when they're exposed to new data, new feedback, and changing conditions. Without regular updates, they stagnate — or worse, drift into inaccuracy.

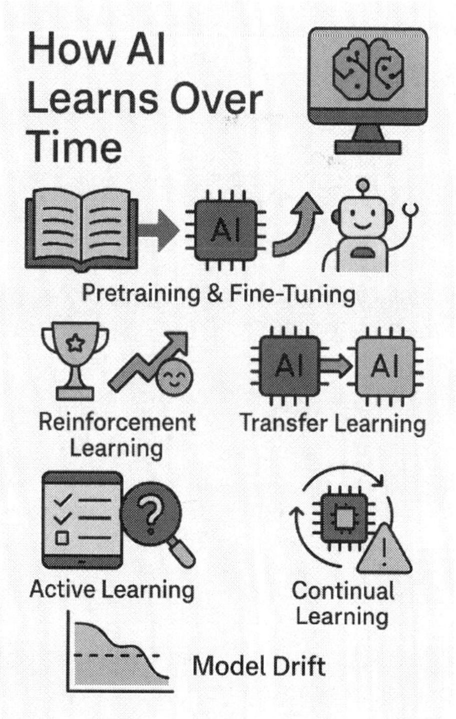

Why Ongoing Learning Matters

The world shifts constantly:

- New slang, trends, or product names appear.
- Customer preferences evolve.
- Market and social conditions change.
- Errors or biases that were once hidden become more visible.

A model trained six months ago may no longer perform well today. This is especially true in fast-moving fields like finance, healthcare, search, or customer support. Without regular updates, AI can become outdated or misaligned with user needs.

Maintenance Is Part of the Job

Ongoing learning doesn't always require retraining a model from scratch. It can take several forms:

Feedback loops:

Systems log corrections or user interactions and use them to improve future outputs.

Scheduled retraining:

The model is periodically updated with new data to reflect emerging trends or edge cases.

Monitoring for drift:

Automated tools detect when accuracy declines or input data begins to shift.

Human feedback:

Expert review, customer input, or user corrections act as valuable teaching signals.

AI systems need care and feeding — not just code and data. They perform best when treated like evolving systems: continuously observed, tuned, and improved.

From Training to Teaching

What we call "machine learning" could just as easily be called **machine teaching**. It's fundamentally a human-guided process — choosing datasets, designing feedback, determining updates, and embedding values. Even the most advanced models still rely on human judgment to steer their learning.

By treating AI as an ongoing learning partner — not a one-off solution — we build tools that are smarter, safer, and more adaptive. The goal is not perfection at deployment, but continuous alignment with the world and the people who use them.

Quick Takeaways: How AI Learns Over Time

» Pretraining vs. Fine-Tuning

AI models start with broad pretraining on general data, then specialize through fine-tuning on domain-specific examples.

» Reinforcement Learning

Some systems learn through feedback loops, improving performance by receiving rewards or penalties for actions — like teaching a dog tricks through trial and error.

» Transfer and Few-Shot Learning

Modern models can transfer knowledge between tasks, and sometimes learn new skills from very few examples — enabling rapid adaptation.

» Continual Learning and Model Drift

AI needs to keep learning to stay accurate. Without updates, performance can degrade over time due to shifting data or changing environments.

» Active Learning

Rather than learning from all data equally, active learning lets models query or prioritize the most informative examples — boosting efficiency.

» AI Needs Maintenance

AI learning isn't a one-time event. Like any system, it requires monitoring, feedback, and ongoing care to remain useful and aligned.

Chapter Summary

This chapter explored how AI learns — and, more importantly, how it keeps learning after its initial training. We began with pretraining, which gives models broad general knowledge, and fine-tuning, which adapts that knowledge to specific tasks or industries. We then looked at how reinforcement learning allows systems to improve through feedback and iterative reward signals.

From there, we examined how **transfer learning, few-shot learning**, and **zero-shot learning** enable modern models to adapt quickly and perform tasks with minimal data — an ability that dramatically lowers the barrier to experimentation and innovation.

Next came the challenge of **continual learning** and why real-world systems must evolve without overwriting what they already know. We discussed the problem of **catastrophic forgetting**, the techniques designed to combat it, and the broader importance of AI systems that can adapt safely and responsibly over time.

We also explored **active learning**, where AI seeks out the most informative data to label — mirroring the way humans focus on areas where they feel least confident.

Finally, we emphasized that deploying an AI model is not the finish line. Real-world AI requires **monitoring, retraining, feedback loops, and human oversight** to remain accurate and aligned. AI is not a static artifact — it's an evolving system shaped continuously by data, context, and the people who guide it.

Learning isn't a moment. It's a cycle — and the most reliable systems are the ones designed with this in mind.

Chapter 14

CAN WE TRUST IT?
EXPLAINABILITY, BIAS, AND HUMAN OVERSIGHT

Introduction: Power Requires Responsibility

Artificial intelligence has become deeply embedded in daily life — from recommendations and voice assistants to hiring tools and healthcare systems. But with this power comes a crucial question: **Can we trust the systems making decisions on our behalf?**

Trust in AI isn't just about performance. It's about whether the system treats people fairly, whether its decisions can be understood, and whether the humans deploying it remain accountable when things go wrong.

In this chapter, we move beyond broad ethical themes and focus on the **practical principles** that determine whether AI can be used responsibly in real-world settings: explainability, transparency, human oversight, bias mitigation, fairness, safety, accountability, and regulation. These are not abstract ideals — they're the guardrails that shape whether AI helps society or harms it.

---

1. Explainability and Transparency

For AI to be trustworthy, it must be understandable. **Explainability** refers to how well a system can justify or clarify its output. Transparency means making clear how the system works, what data it uses, and how decisions are made.

When people are affected by AI decisions — whether they're applying for a loan or being flagged by a facial recognition system — they deserve to know why. A "black-box" system may produce results, but if the reasoning is unclear or hidden, trust erodes quickly, especially in high-stakes domains.

Real-world example:

Clearview AI faced global backlash for scraping billions of images from social media without consent. Its opaque methods raised alarms about surveillance, privacy, and user control — demonstrating how a lack of transparency can undermine public confidence.

Best Practices Include:

- **Model documentation** (e.g., datasheets or model cards)
- **Visual or natural-language explanations** of decisions
- **Clear terms of use and privacy disclosures**

Transparency is not just a technical feature — it's a public right.

Why Explainability Matters

- Users deserve to understand how decisions are made.
- Regulators may require justification for automated decisions (e.g., loan approvals, medical diagnostics).
- Engineers need visibility into the system's reasoning to debug, refine, and improve performance.

Types of Explainability

Intrinsic Explainability

Some models — such as decision trees or linear regressions — are inherently interpretable. Their structure makes it easy to see how inputs map to outputs.

Post-Hoc Explainability

Techniques like LIME or SHAP generate explanations after the fact, even for opaque deep-learning models.

Visual Explanations

In computer vision, attention maps or heatmaps highlight the parts of an image that influenced a model's decision.

Explainability is ultimately about more than transparency — it's about communicability. An explanation is only useful if it is understandable and tailored to the needs of different stakeholders, from end-users to regulators to technical teams.

---

2. Human-in-the-Loop AI: Keeping People in Control

AI can analyze, predict, and automate — but full autonomy is not always desirable or safe. In many contexts, decisions carry consequences that demand human judgment. Human-in-the-Loop (HITL) systems address this by keeping people actively involved in the decision cycle.

In HITL systems, AI may flag, recommend, or predict — but **humans remain responsible for the final call.** This ensures accountability, ethical grounding, and resilience when real-world complexity exceeds what algorithms can reliably handle.

Why HITL Matters

The goal of HITL is not to slow AI down, but to keep it aligned with human values, especially where consequences are significant.

Benefits of HITL

- Provides a fail-safe for high-stakes decisions
- Allows real-time correction of model errors
- Ensures ethical oversight where rules alone fall short
- Improves systems over time through human feedback
- Maintains accountability and reduces risk of blind automation

HITL is a reminder that AI augments humans — it doesn't replace them.

Critical Use Cases

HITL is essential in sectors where precision, fairness, and accountability matter:

Healthcare

AI can highlight patterns in scans or predict outcomes, but doctors must validate results. A missed nuance could cost a life.

Law Enforcement and Criminal Justice

Predictive tools may identify trends, but relying solely on algorithms risks reinforcing historical biases. Humans must make the actual decisions.

Military and Defense

Autonomous weapons raise urgent ethical and legal concerns. International norms insist lethal actions remain under human control.

Content Moderation

AI flags harmful content, but human reviewers provide cultural context, nuance, and judgment — especially around sensitive speech.

Financial Systems

Fraud detectors or credit algorithms may raise alerts, but human auditors review edge cases, catch false positives, and ensure equitable treatment.

What Makes a System "Human-in-the-Loop"?

HITL is not simply placing a human nearby — it's about **designing systems that expect, support, and rely on human involvement.**

Key HITL Design Features

- **Intervention interfaces:** Clear, intuitive ways for humans to override or modify AI outputs
- **Audit trails:** Logs that record every interaction for transparency and accountability
- **Feedback loops:** Human corrections used to improve model behavior
- **Training and communication:** Helping users interpret AI recommendations and understand their limits

When built properly, HITL turns AI from an autonomous operator into a collaborative partner.

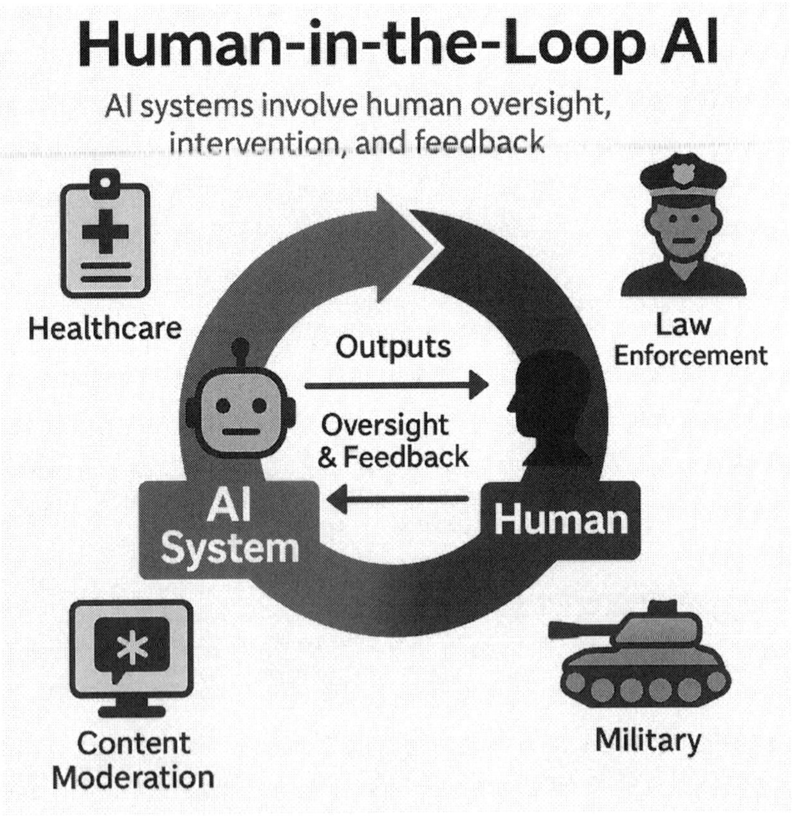

HITL vs. Full Automation

HITL is often misunderstood as a limitation. In reality:

- **Automation offers speed**, but increases the risk of unchecked errors.
- **HITL offers balance**, pairing machine efficiency with human insight.
- **Shared responsibility** reduces harm and improves trust.

Where stakes are high, HITL is not optional — it is the foundation of responsible AI.

---

3. Algorithmic Bias: Where It Comes From

AI is often described as objective — a neutral decision-maker. But AI systems learn from data, and data reflects society. If society is biased, the **model will learn those biases**, replicate them, and sometimes amplify them.

A well-known example is Amazon's hiring tool (2014–2015). Trained on resumes from a predominantly male workforce, the model learned to penalize terms associated with women and favor those associated with men. Despite attempts to fix it, the system continued producing biased recommendations and was ultimately abandoned. The problem wasn't a technical bug — **it was baked into the data.**

Bias isn't a rare exception in AI. It's a predictable outcome unless it is actively addressed.

Common Sources of Algorithmic Bias

Bias can enter AI systems from several directions. These are the most frequent — and the most important to understand:

Historical Bias

Patterns of inequality encoded in past data are learned and repeated by AI.

Example: arrest records or hiring history reflecting systemic discrimination.

Sampling Bias

When some groups are underrepresented in the data.

Example: facial recognition systems performing poorly on darker skin tones because datasets skewed light.

Measurement Bias

When the features chosen to represent an idea are flawed.

Example: using zip code as a proxy for creditworthiness embeds socioeconomic bias.

Labeling Bias

Human annotators introduce subjective judgments.

Example: inconsistent labeling of emotion, toxicity, or "professionalism" based on personal bias.

Feedback Loops

AI predictions shape future data in ways that reinforce bias.

Example: predictive policing sends officers to the same neighborhoods, creating more arrest data and "proving" the system right.

Bias isn't an abstract risk — it's a structural one.

Consequences of Bias

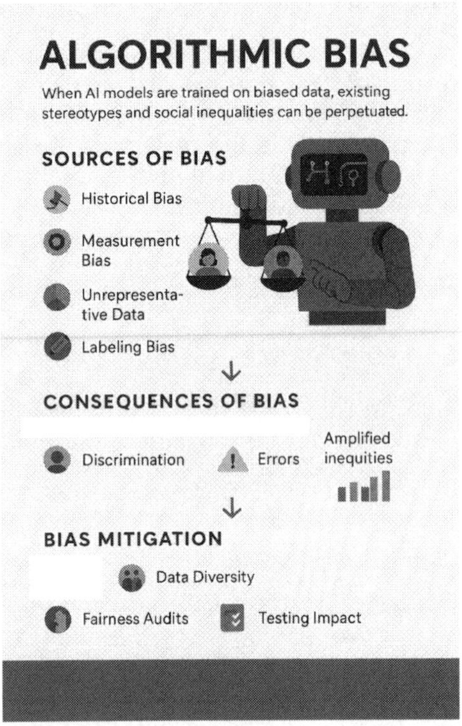

Bias in artificial intelligence isn't a theoretical risk — it has concrete, often harmful consequences that ripple through society. When AI systems inherit or amplify societal prejudices, the result can be unjust outcomes in deeply consequential areas such as hiring, lending, housing, healthcare, and criminal justice. A biased hiring algorithm may overlook qualified candidates from underrepresented backgrounds simply because it was trained on historical data favoring a homogenous workforce. A credit scoring system might penalize individuals based on proxy variables like zip codes, reinforcing patterns of socioeconomic inequality.

These outcomes are not merely technical errors — they represent a breakdown in fairness and accountability, undermining the legitimacy of AI as a decision-making tool. When people begin to see AI as biased or opaque, trust erodes — both in the technology and in the institutions that deploy it. That erosion can lead to public backlash, legal challenges, and reputational harm for companies, especially when the biases reflect systemic discrimination that society is actively working to undo. If unaddressed, biased AI systems don't just reflect inequality — they deepen it, embedding unjust assumptions into decisions that affect millions.

Bias is not a glitch; it is a structural force that shapes outcomes. Left unchecked, it reinforces the very inequalities society is trying to overcome.

Strategies for Bias Mitigation

Eliminating bias entirely may not always be possible, but confronting and mitigating it is not optional — it's a moral and operational imperative. Addressing bias requires careful, continuous intervention throughout the AI lifecycle.

It begins with the data: training datasets must be broad, diverse, and representative of the populations the system will serve. Otherwise, the model risks excluding or misrepresenting key groups from the outset. Rigorous testing is essential — fairness audits and impact assessments should be built into development workflows, with explicit checks for how the system performs across lines of race, gender, age, ability, and other identity markers. But data and metrics alone aren't enough. The people designing, building, and evaluating AI systems must also reflect a diversity of experiences and perspectives, and teams must be transparent about how models were built and where their limitations lie.

To turn these principles into action, organizations should:

1. Build Better, More Representative Datasets

- Collect data that reflects the diversity of real-world populations
- Audit for underrepresentation of key groups
- Avoid proxy variables (like zip code) that encode hidden social and economic bias

2. Perform Systematic Fairness Testing

- Measure error rates across demographic groups
- Use fairness metrics such as demographic parity or equalized odds
- Conduct pre-deployment impact assessments to surface disparities early

3. Design with Diverse Teams

Homogeneous teams miss blind spots. A mix of backgrounds, identities, and expertise helps anticipate harms and challenge unexamined assumptions.

4. Prioritize Transparency and Documentation

Model cards, data sheets, and impact reports make the model's training and limitations clear — enabling accountability and informed oversight.

5. Monitor Continuously After Deployment

Bias can emerge over time as data shifts. Models need ongoing audits, feedback loops, and revision cycles—not a one-time fairness check.

Bias mitigation isn't a box to tick. It's an ongoing commitment to care, correction, and responsibility — and the foundation of trustworthy AI.

Why It Matters

Bias in AI is not a side note. It is one of the defining ethical challenges of our century.

At stake is more than accuracy — it's fairness, dignity, and equal opportunity.

When we take bias seriously — through inclusive design, transparent documentation, and proactive correction — AI becomes not just a powerful tool, but an instrument for more equitable decision-making.

It's not about striving for perfection. It's about choosing responsibility, again and again.

---

4. Fairness, Safety and Accountability

Trustworthy AI is not just about accuracy or performance — it's about ethics. For an AI system to earn genuine trust, it must be fair in how it treats people, safe in how it operates, and **accountable** when things go wrong. These aren't optional extras; they are foundational pillars of responsible AI.

Fairness: Avoiding Disparate Impact

Fairness means ensuring that AI systems do not disproportionately harm, exclude, or misrepresent individuals based on attributes such as race, gender, age, socioeconomic status, disability, or any other protected category.

But fairness is complex — there is no single universal definition. Different contexts require different fairness criteria. Common frameworks include:

- **Demographic parity:** Outcomes are statistically similar across groups.
- **Equal opportunity:** Error rates (e.g., false positives or negatives) are balanced for all groups.
- **Counterfactual fairness:** Changing a sensitive attribute (like gender) shouldn't change the model's decision if all else remains equal.

Example:

If an AI hiring tool selects fewer qualified women than men due to skewed training data, that is not a "technical mistake" — it is a fairness failure, and it must be corrected.

Fairness is ultimately about equitable outcomes, not just mathematical definitions.

Safety: Building Systems That Don't Harm

AI systems can cause real harm when they fail — especially in high-stakes areas like healthcare, transportation, finance, or public safety. Safety means minimizing unintended consequences and ensuring systems behave reliably, even in unpredictable conditions.

Key safety practices include:

- Rigorous testing in realistic, edge-case-rich environments
- Red-teaming to probe vulnerabilities and rare failure modes
- Fail-safes and human override mechanisms for critical decisions

A misread X-ray, a mistaken stop-sign classification, or an incorrect medication recommendation can lead to catastrophic outcomes. Safety demands thoroughness, not assumptions.

Accountability: Knowing Who's Responsible

When AI makes (or influences) a decision, someone must be responsible.

A denied loan, an unfair profile, or a mistaken medical prediction cannot be blamed on "the algorithm."

To enforce accountability, organizations must:

- **Maintain audit trails** of model inputs, decisions, and outcomes
- **Assign ownership** of AI systems and their impacts within the organization
- **Provide redress mechanisms** so users can contest or understand decisions

- **Ensure model governance** so that changes, updates, and rollbacks are traceable

Accountability ensures that AI is not a free-floating decision-maker without oversight—it is part of a governed system.

Turning Ethics into Action

Ethical AI requires deliberate, ongoing effort.

Intentions are not enough.

Organizations should commit to:

- **Including diverse perspectives** during model design
- **Using ethical review boards** for oversight of high-risk systems
- **Continuous monitoring** as data and conditions evolve
- **Documenting limitations and risks** before deployment

Fairness isn't perfection.

Safety isn't immunity.

Accountability isn't blame.

They are commitments — to care, to vigilance, and to keeping people at the center of every AI decision.

---

5. Responsible AI and Regulation

Building AI responsibly requires more than good intentions. It demands **governance frameworks, clear policies**, and increasingly, **formal regulation**. As AI systems shape decisions in hiring, healthcare, finance, education, and policing, society expects guardrails that ensure these tools are used ethically, safely, and transparently.

From Principles to Practice

Responsible AI begins with values — fairness, privacy, safety, transparency — but values must be translated into action. That means designing systems with ethics embedded from the start, not patched in afterward.

Common operational strategies include:

Privacy-by-Design

Building systems that respect data rights and confidentiality at every stage.

Impact Assessments

Evaluating potential harms, biases, and unintended consequences before deployment — especially in high-risk domains.

Sustainability Considerations

Understanding and minimizing the environmental cost of training and running large AI models.

Documentation and Disclosure

Using tools like model cards, datasheets, and risk reports to communicate what data was used, how the model was trained, where it performs well, and where it does not.

Documentation isn't bureaucracy — it's how organizations make invisible processes visible, which is essential for trust.

Regulatory Landscape

Governments and institutions worldwide are beginning to design laws and frameworks that set expectations around transparency, risk, and accountability.

Key examples include:

GDPR (European Union)

Requires user consent for data use and gives citizens rights over automated decision-making, including explanations.

CCPA (California Consumer Privacy Act)

Grants consumers control over what data companies collect, store, and sell.

EU AI Act

One of the most comprehensive AI regulations to date.

It classifies AI systems into risk levels — from minimal to unacceptable — and mandates:

- Human oversight for high-risk systems
- Traceability and record-keeping
- Strict transparency obligations
- Quality and governance of training data

NIST AI Risk Management Framework (U.S.)

A voluntary but influential framework guiding organizations on building trustworthy, safe, and secure AI systems.

These frameworks share a common goal: ensuring AI aligns with human rights, democratic values, and public trust — especially in sensitive or high-impact settings.

As discussed in the RAG chapter, transparency becomes even more important when generative systems pull external information into their answers; fact-grounding through retrieval makes outputs auditable and reduces risk.

Why Ongoing Oversight Matters

AI is not "set it and forget it." Even a high-performing model can degrade over time due to drift, new behaviors in the data, or evolving user expectations. Responsible AI requires continuous oversight.

This includes:

- **Monitoring performance** in real-world use
- **Auditing outputs** for bias or harmful patterns
- **Updating or retracting models** when necessary
- **Clear lines of accountability** for every system deployed
- **Defined user recourse** when decisions cause harm

Without ongoing governance, even well-intentioned models can become opaque, unfair, or dangerous as the world shifts around them.

Regulation and responsible AI frameworks ensure that organizations remain accountable for the systems they build — from development to deployment to retirement.

---

6. Sector-Specific Ethics: Context Matters

AI systems do not operate in isolation. They function inside real-world domains — each with its own sensitivities, risks, regulations, and expectations. What is ethically acceptable in marketing, for example, may be unacceptable in medicine or criminal justice.

Ethical AI must be contextual. Principles like fairness, transparency, and safety remain constant, but how they are applied varies dramatically by sector.

Healthcare

AI in healthcare can assist with diagnosis, triage, treatment recommendations, and operational efficiency — but mistakes can have life-or-death consequences. Because of this, ethical use in healthcare demands heightened scrutiny.

Key considerations include:

- **Rigorous clinical validation** with medical experts before deployment
- **Protection of sensitive health data** under laws like HIPAA and GDPR

- **Explainability requirements** so clinicians understand how AI reached a recommendation
- **Human-in-the-loop oversight**, ensuring clinicians remain responsible for final decisions
- **Bias monitoring**, since uneven training data can lead to misdiagnosis in underrepresented groups

In healthcare, AI must enhance — not replace — expert judgment. Trust is built only when safety, transparency, and patient autonomy are prioritized.

Finance

AI systems in finance influence credit scores, loan approvals, fraud detection, and investment decisions. Ethical failure here can reinforce inequality or cause widespread economic harm.

Core principles include:

- **Auditable decision-making** for credit, lending, and insurance
- **Fair treatment regardless of race, gender, age, or zip code**
- **Clear explanations** for automated denials or risk assessments
- **Protection against proxy discrimination**, where seemingly neutral variables encode systemic bias
- **Human review mechanisms** for disputed or unclear outcomes

Because financial decisions materially affect people's lives, fairness and accountability are not optional — they are regulatory and ethical imperatives.

Law Enforcement and Security

No domain combines as much risk, scrutiny, and controversy as policing, surveillance, and security. Tools like facial recognition, predictive policing, biometric monitoring, and real-time tracking raise profound ethical concerns.

Responsible use requires:

- **Transparency** in how systems are trained, deployed, and evaluated

- **Strict legal limits** to prevent mass surveillance or abusive monitoring
- **Oversight bodies** to enforce civil liberties and human rights
- **Bias auditing**, especially for face recognition models that historically underperform on darker skin tones
- **Clear accountability** for errors that lead to wrongful arrests or investigations

Given the potential for irreversible harm, law enforcement AI demands the highest standard of governance.

Education

AI is increasingly used for grading, admissions, adaptive learning, tutoring, plagiarism detection, and behavioral analytics. Ethical use in education must prioritize equity and student wellbeing.

Important considerations include:

- **Transparency** in how assessments or recommendations are generated
- **Avoiding automated decisions** that reinforce socioeconomic or racial disparities
- **Equitable access** to AI-powered tools so they don't privilege already-advantaged students
- **Privacy safeguards**, especially for minors and sensitive academic data
- **Human oversight**, ensuring teachers retain control and interpret AI suggestions in context

In education, AI should expand opportunity and personalize learning — not entrench privilege or outsource judgment.

Why Sector Context Matters

AI ethics is not a universal checklist. It is a **context-dependent practice** shaped by domain-specific risks, users, laws, and consequences.

A model safe for entertainment may be dangerous in healthcare.

A system acceptable for marketing may be unlawful in law enforcement.

Responsible AI development means:

- recognizing the environment in which a system operates,
- tailoring safeguards to that environment, and
- engaging the relevant domain experts early and continuously.

Ethical AI isn't just about what a system can do — it's about what it should do in each specific context.

7. Access and Equity: Who Gets to Benefit?

AI has extraordinary potential to uplift societies — but it can just as easily deepen global inequality. When access to AI tools, data, infrastructure, and education is limited to a small group of wealthy nations, large corporations, or highly technical users, the digital divide widens. Equity is not just a moral concern; it's a technical, economic, and societal imperative.

The Risk of Widening Gaps

Without intentional intervention, AI may:

- **Concentrate power** among a handful of companies with the resources to train massive models
- **Reinforce dominant languages and cultures**, marginalizing smaller or Indigenous ones
- **Exclude communities without high-bandwidth internet or modern devices**
- **Place certain workers at higher risk of job displacement**, without offering pathways to reskill
- **Overlook local contexts** — producing systems that work well in Silicon Valley but fail in rural, low-resource, or culturally diverse environments

This creates a self-perpetuating cycle: those with access to AI reap the benefits (efficiency, automation, insight), while those without fall further behind.

Responsible AI Includes Everyone

A fair AI ecosystem requires deliberate design choices and investments that expand access and representation:

1. Education and Digital Literacy

Equitable AI begins with empowering people to understand and use the technology.

Examples include:

- Accessible AI education programs in schools and community centers
- Local-language curricula
- Free or low-cost online courses that emphasize practical, real-world skills
- Training pathways for workers adapting to automation

When communities understand the technology, they can shape it — not just consume it.

2. Open Source and Publicly Available Models

Proprietary AI tools are expensive and closed by design. Open-source models and datasets reduce barriers to innovation by enabling:

- Small businesses and startups to experiment without prohibitive costs
- Local developers to build region-specific tools
- Researchers in low-resource regions to contribute meaningfully to AI advancement

Initiatives like open-weight language models, public datasets, and community-led research hubs democratize participation.

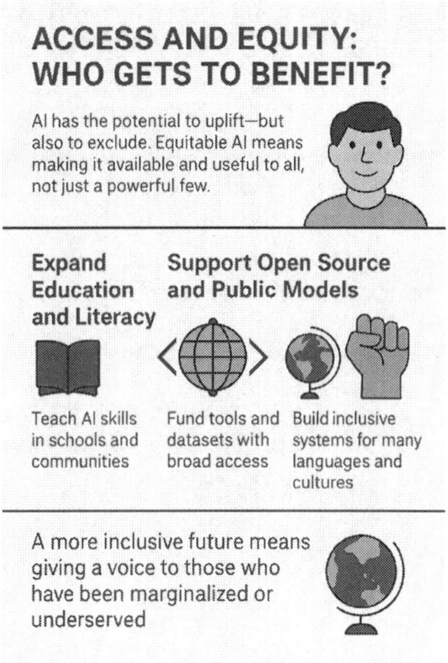

3. Designing for Diverse Contexts

AI should work for everyone — including users with disabilities, non-dominant languages, rural communities, and culturally diverse populations.

This means:

- Multilingual and multimodal support
- Localization and culturally aware datasets
- Accessibility-first design (screen readers, voice interfaces, simplified UI modes)
- Tools that operate on low-bandwidth devices or offline

Good design anticipates diversity, not uniformity.

4. Community Participation and Co-Design

Ethical deployment cannot be top-down.

Meaningful inclusion requires:

- Engaging communities early in system design
- Understanding local values, challenges, and expectations
- Creating feedback loops that allow users to shape system behavior
- Ensuring AI tools respect cultural norms and community priorities

AI designed with communities — not just for them — earns trust and stays relevant.

A More Inclusive Future

Access isn't just about who uses AI — it's about who builds it, who benefits from it, and who has a voice in shaping its direction. Equity in AI means building systems that reflect humanity in all its diversity, and ensuring no one is left behind in the next wave of innovation.

---

8. Wrapping Up: Ethics Is Ongoing

Ethics in AI isn't a box to check — it's an evolving responsibility. Technology changes, societies change, and expectations change. What is considered responsible today may require reevaluation tomorrow. That adaptability is the heart of trustworthy AI.

A reliable AI system must be:

- **Transparent** — so people understand what it does and why
- **Fair** — so it treats individuals and groups equitably
- **Safe** — so it minimizes harm and behaves reliably
- **Accountable** — so someone is responsible for outcomes
- **Inclusive** — so its benefits are shared broadly, not concentrated among a few

But above all, responsible AI keeps humans at the center. Machines may optimize, predict, or generate — but humans define the values, set the boundaries, and make the final judgments.

As AI grows more powerful and pervasive, trust doesn't come from capability alone. It comes from the care we put into designing, governing, and monitoring these systems over time. Ethical AI isn't just about preventing failure; it's about earning confidence — day after day, decision after decision.

Quick Takeaways: Can We Trust It?

» Explainability

AI decisions must be understandable. Clear explanations and transparent documentation help users, engineers, and regulators grasp how and why a system arrived at an outcome.

» Human-in-the-Loop (HITL)

Keeping people involved in key decisions ensures oversight, ethical judgment, and correction — especially in high-stakes domains like healthcare, finance, and law enforcement.

» Bias and Fairness

AI inherits patterns from data. Without deliberate checks, this can lead to unfair or harmful outcomes. Mitigating bias requires diverse data, inclusive teams, transparency tools, and continuous auditing.

» Accountability by Design

Trustworthy AI demands clear ownership, audit trails, and mechanisms for appeal or redress. Responsibility cannot be outsourced to the algorithm.

» Governance and Regulation

Frameworks like GDPR, the EU AI Act, and the NIST RMF provide guardrails for privacy, transparency, and safe deployment. Regulation is becoming integral to responsible AI practice.

» Access and Equity

AI should benefit everyone — not just those with resources or technical expertise. Inclusive datasets, open resources, and global digital literacy initiatives help reduce inequity.

Chapter Summary

Trust in AI isn't measured only by performance — it's measured by how systems behave, whom they affect, and whether their decision-making can be explained and governed responsibly. This chapter explored the foundational elements of trustworthy AI: explainability, transparency, and the role of human oversight. It examined how bias emerges from historical and structural patterns, and how fairness requires intentional design, diverse teams, and continuous monitoring.

We explored why high-stakes decisions must keep humans firmly in the loop and how governance frameworks — from GDPR to the EU AI Act — shape the safe and ethical deployment of AI. Sector-specific ethics highlighted that what's "responsible" varies across contexts such as healthcare, finance, education, and law enforcement. Finally, we emphasized the importance of access and equity to ensure that the benefits of AI are broadly shared, not concentrated.

The takeaway is clear: ethical AI is not a one-time achievement. It is an ongoing commitment to transparency, fairness, safety, accountability, and inclusion — a commitment that grows more important as AI becomes more capable and more deeply embedded in society.

Part III AI in Everyday Life

Introduction: Living with AI

Artificial intelligence is no longer confined to labs, research papers, or specialized teams. It's woven into everyday life.

By now, you've seen how AI works under the hood — how models are trained, how they adapt, and how we keep them fair, safe, and accountable. That's the technical foundation.

But most people don't experience AI as algorithms or training data. They experience it through everyday moments.

The real questions become:

Where is AI already in my life?

And how can I use it better, more intentionally, and more safely?

This part of the book zooms out from model architecture and data pipelines to the world you live in — your phone, your home, your workplace, your creative tools, and even your daily routines. Whether you notice it or not, AI already shapes how you find recipes, choose movies, write emails, track your sleep, shop online, navigate traffic, or plan your week.

You don't need to be a programmer to use AI.

But understanding where it shows up — and what it's doing behind the scenes — helps you make smarter choices, avoid pitfalls, and unlock real value in your everyday tools.

In the chapters ahead, we'll explore:

- AI in your phone and personal assistants
- Smart homes that automate comfort, safety, and energy use
- AI at work — from productivity to collaboration and decision-making
- Creative tools that help you write, design, edit, and compose
- Healthcare technologies that support diagnosis, wellness, and prevention
- Self-driving cars and the evolving future of personal mobility

None of this is speculative or futuristic — it's already part of daily life.

Understanding these systems empowers you to get more out of them, to recognize their limitations, and to decide where AI should assist you versus where it shouldn't.

Whether your goal is to save time, stay healthy, unlock creativity, improve your work, or simply understand the world as it changes, AI can be a powerful ally — if you know how to use it wisely, critically, and intentionally.

Chapter 15

AI IN YOUR POCKET HOW SMARTPHONES AND WEARABLES MAKE LIFE SMARTER

Introduction: The Smartest Thing You Own

Most people think of AI as something futuristic — but chances are, you're carrying one of the most powerful AI-enabled devices in your life right now. Your smartphone sees, hears, predicts, translates, filters, organizes, and personalizes far more than you might realize. It's not just a communication tool; it's a pocket-sized AI platform that works constantly behind the scenes.

In this chapter, we explore the many ways AI operates quietly (and sometimes loudly) inside your phone and wearable devices — shaping your daily routines, simplifying tasks, and anticipating your needs. More importantly, you'll learn how to use these systems intentionally, so AI becomes a helpful companion rather than background noise.

---

1. AI Features in Modern Smartphones

Today's smartphones are packed with AI-powered features that run continuously in the background or activate with a single tap. Many of the tools you rely on every day — sometimes without noticing — are driven by machine learning, computer vision, or natural language processing.

Voice Assistants (Alexa, Siri, Google Assistant, Bixby)

These assistants use speech recognition and NLP to interpret your requests and perform tasks such as:

- Answering questions
- Setting reminders
- Managing your smart home

- Sending messages
- Delivering news, weather, or directions

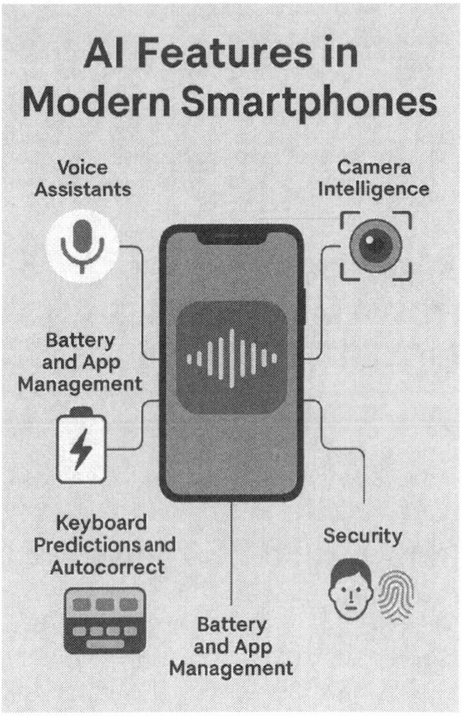

A report published by Statista indicates that the number of digital voice assistants in active use has doubled in just five years — from 4.2 billion to 8.4 billion — underscoring how essential they've become for everyday interaction.

(We first introduced voice assistants in Chapter 4, and we'll revisit their evolution into generative AI models in Chapter 16.)

Camera Intelligence

Modern cameras are essentially small AI systems:

- Detecting scenes and adjusting settings
- Enhancing lighting and sharpness
- Stabilizing shaky footage
- Suggesting the best moment to capture a photo
- Offering real-time filters and optimization

This is why even mid-range phones now deliver photos that once required professional equipment.

Battery and App Management

Your phone learns your habits — when you open certain apps, which ones drain power, and what can safely be paused. AI uses these patterns to:

- Optimize battery usage
- Freeze or limit background tasks
- Predict when you'll need power and adjust accordingly

Keyboard Predictions and Autocorrect

AI-powered language models anticipate what you're about to type, fix typos, and learn your writing style over time.

This is why your phone might complete your sentences — sometimes eerily well.

Security Features

Biometric systems like fingerprint scanning and face recognition rely on computer vision to unlock your phone quickly and securely.

These models compare your features in milliseconds, often entirely on-device for privacy and speed.

Example: Google Pixel

Google's Pixel phones lean heavily on on-device AI to:

- Screen suspicious calls
- Identify music playing nearby
- Summarize voicemails
- Enhance photos using advanced machine learning

All while minimizing cloud use to protect your data.

---

2. Everyday Apps Powered by AI

AI isn't limited to your phone's built-in features — it also powers the apps you use every day. Many of these apps run quietly in the background, learning your habits and tailoring themselves to you. What once required large servers is now done through lightweight, efficient machine-learning models integrated directly into everyday tools.

Maps and Navigation

AI predicts traffic patterns, analyzes real-time conditions, suggests alternate routes, and even estimates the best time to leave. Apps like Google Maps and Waze use millions of data points — speed, congestion, accidents, and historical trends — to keep you moving efficiently.

Photos

Modern photo apps use AI to recognize faces, objects, locations, and even emotions. This makes searching effortless ("birthday," "dog," "beach"), and allows automatic organization into albums, memories, or highlights. AI also assists with enhancements such as sharpening, color correction, or object removal.

Streaming Platforms

Netflix, YouTube, Spotify, and other platforms rely heavily on recommendation models. They study viewing or listening habits, compare them to similar users, and surface content you're more likely to enjoy. This personalization is why your feed rarely looks like someone else's — even if you share the same household.

Fitness Trackers and Health Apps

Connected wearables and phone sensors generate data about sleep patterns, steps, heart rate, movement, and activity. AI interprets this information to provide actionable insights: when to move, how well you slept, or whether your heart rate looks unusual.

Language and Translation Apps

Tools like Duolingo and Google Translate use AI to adapt lessons to your skill level, fix pronunciation, predict mistakes, and translate text, audio, or images in real time. Behind the scenes, language models track your progress the same way a personal tutor would.

Shopping and Product Discovery

Retail apps use machine learning for personalized recommendations, search ranking, dynamic pricing, and visual search (upload a photo → find similar products). Some apps even predict what you might reorder or need soon, based on past behavior.

Messaging and Typing

AI-driven suggestions and grammar correction have evolved far beyond simple autocorrect. Modern language models understand tone, context, and intent — enabling features like smart replies, multilingual typing, or predictive text that adapts to your personal writing style.

These apps aren't just tools — they're personalized AI companions that learn from your behavior and help streamline daily routines. The more you use them, the better they get at anticipating your needs.

---

3. On-Device vs. Cloud AI

AI on your smartphone works through two main pathways: on-device processing (local) and cloud-based processing (remote). Both are essential, but they serve different purposes — and understanding the difference helps you make better choices about performance, privacy, and data use.

On-Device AI

On-device AI means the model runs directly on your smartphone's hardware. Thanks to specialized chips — Apple's Neural Engine, Google's Tensor processors, and Qualcomm's AI accelerators — phones can now perform tasks that once required large servers.

Key advantages include:

- **Speed and low latency:** responses are instantaneous, ideal for real-time tasks like face unlock, live transcription, or camera stabilization.
- **Offline capability:** features continue to work even without an internet connection.
- **Improved privacy:** data stays on your device instead of being sent to external servers.
- **Battery efficiency:** modern chips optimize power usage by minimizing network calls.

Examples of on-device AI:

- Face ID / Face Unlock
- Live speech captioning
- Photo processing (night mode, portrait blur, scene detection)
- Music recognition in the background (e.g., Pixel's "Now Playing")
- Predictive keyboard suggestions trained on your local typing history

On-device AI turns your phone into a self-contained intelligence module, reducing dependence on cloud networks and protecting personal data by design.

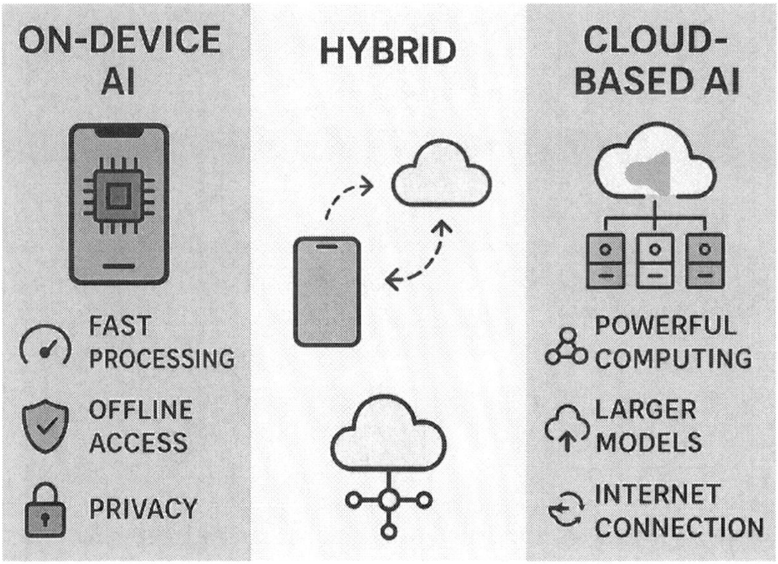

Cloud-Based AI

Cloud AI relies on external servers — massive data centers running heavyweight models. These systems handle tasks too large or too complex for a smartphone to process locally.

Why cloud AI matters:

- **Much greater computing power:** ideal for large language models (like ChatGPT or Gemini), advanced translation, or deep search.
- **Access to fresh, real-time information:** weather, traffic, news, search queries.
- **Shared intelligence:** improvements made on cloud models apply instantly to all users.
- **Support for massive datasets:** cloud services can store, index, and analyze information far beyond a phone's storage.

Examples of cloud AI:

- ChatGPT, Gemini, Claude
- Google Photos cloud backup and search
- Real-time language translation
- Large-scale spam detection
- Smart email categorization
- Cloud-based smart assistants

Cloud computing, broadly, means using remote servers to store data, run software, or perform computation. In AI terms, it means your device captures the input, but the analysis and heavy lifting happen in a remote data center — giving you access to supercomputer-level intelligence.

Privacy Tip: Local AI When Possible

Many modern devices now offer local processing for sensitive tasks like speech transcription, photo classification, or biometric recognition. This protects privacy by avoiding unnecessary data transmission.

You can often enable or prioritize on-device processing in your Privacy or App Permissions settings.

---

4. How to Use Smartphone AI Intentionally

You don't need to be technical to get real value from the AI already inside your phone. Most people use only a fraction of what's available — not because the features are hidden, but because they've never explored them with intention. With a few small habits, you can turn your phone into a personalized assistant rather than a passive tool.

Customize Voice Routines

Voice assistants can do far more than answer questions. You can create multi-step routines activated by a single phrase.

Examples:

- **"Good morning"** → read the weather, play news, show your agenda
- **"Heading home"** → send an ETA text, start navigation, adjust lights
- **"Work mode"** → silence notifications, open focus apps, start a playlist

These workflows save time and reduce friction in your day.

Train Your Keyboard

Predictive text and autocorrect become dramatically better when you use them consistently.

Over time, the keyboard learns:

- your tone and writing style
- your frequently used names, places, or jargon
- how you structure sentences

Tip: Instead of fighting autocorrect, correct it once — it will adapt.

Explore Accessibility Tools (Even If You Don't Need Them Yet)

Modern phones include powerful AI-driven accessibility features that everyone can benefit from:

- **Live transcription:** real-time subtitles for conversations or videos
- **Text-to-speech:** have emails, web pages, or books read aloud
- **Screen readers:** AI that interprets visual content verbally
- **Image/scene descriptions:** especially useful for visually impaired users

These features are hidden gems for productivity, reading, learning, and travel.

Search Smarter in Photos

Stop scrolling. AI can search your entire photo library by meaning. Try keywords like:

- "receipts"
- "sunset"

- "dog"
- "birthday"
- "documents"
- "food"

You'll instantly find photos you forgot you even took.

Use Smart Call Features

Modern phones use AI to handle everyday interruptions:

- Call screening: filter spam or unknown callers
- Voicemail summaries: condensed text you can read in seconds
- Real-time captions: for loud environments or accessibility

Enabling these can reduce stress and save time throughout the day.

Automate Routine Tasks

AI assistants can now chain multiple tasks together:

- send messages
- set reminders
- adjust smart lights
- create events
- check your shopping list
- trigger focus modes

The more repetitive the task, the more useful automation becomes.

Pro Tip: Explore Your Settings

Most of the AI features you already paid for are buried in your phone's menus.

In just 10 minutes, you can set up:

- predictive keyboard training
- call screening
- personalized assistant routines
- privacy-friendly local processing
- photo recognition categories
- battery optimization powered by AI

A quick review of your settings can unlock more value than downloading three new apps.

5. Wearables and Companion Devices

Smartphones may be the central hub of your digital life, but wearables extend that intelligence to your wrist, your workouts, and even your sleep. These devices don't just collect data — they translate it into insight, turning everyday patterns into actionable recommendations.

Health Monitoring

Wearables track far more than steps. Modern sensors capture:

- Heart-rate and heart-rate variability
- Sleep stages and sleep quality
- Blood oxygen levels (SpO_2)
- Stress patterns and recovery times
- Menstrual and fertility cycles
- Fall detection or irregular rhythm alerts

This real-time data helps you understand trends rather than guesswork, giving you early signals when something seems off.

Personal Coaching

AI models analyze your movement, sleep, or heart-rate patterns to deliver personalized coaching:

- prompts to stand or breathe
- suggestions to improve sleep consistency
- stress-reduction nudges
- workout recommendations based on recovery

It's like having a mini wellness coach that quietly adapts to your lifestyle.

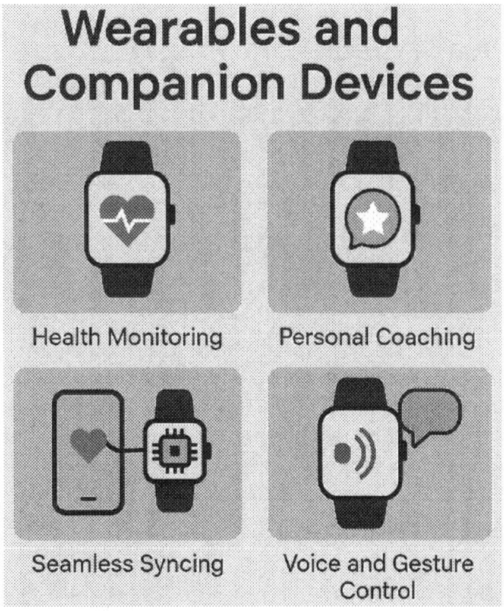

Seamless Syncing With Your Phone

Your wearable feeds data into apps like Apple Health, Google Fit, Samsung Health, or Fitbit — creating a 360° view of your wellbeing.

This unified dashboard helps you:

- track progress over weeks or months
- spot trends that one device alone might miss

- share data with doctors more easily
- understand how sleep, stress, and exercise influence each other

Wearables turn raw data into clearer insights when connected with your phone's ecosystem.

Voice, Gestures, and Hands-Free Control

Many wearables allow hands-free interaction through:

- voice commands (reply to messages, start timers, navigate)
- AI-interpreted gestures to scroll, tap, or answer calls
- quick-reply suggestions generated from context

These features make wearables especially useful during workouts, commuting, or times when your phone isn't accessible.

A Growing Role in Digital Wellness

As wearables evolve, they're becoming less like accessories and more like continuous companions — predicting needs, spotting health irregularities, and helping you form better habits.

With increasingly advanced sensors and on-device AI, the next generation of wearables will be even more proactive and context-aware.

---

6. Risks and Boundaries

AI on your smartphone brings real convenience — but it also introduces real trade-offs. Understanding these risks helps you use your device more intentionally rather than letting it shape your behavior by default.

Over-Personalization: The Echo Chamber Effect

Recommendation systems learn quickly — sometimes too quickly.

AI-powered feeds can:

- reinforce your existing tastes
- filter out diverse viewpoints
- trap you in a loop of similar music, videos, or ideas

What feels convenient can subtly narrow your world if you never reset or diversify your inputs.

Surveillance and Data Monetization

Smartphones and apps collect vast behavioral data — often more than users realize.

Risks include:

- location tracking in the background
- microphone or camera access you didn't intend to grant
- your data being shared or sold to advertisers
- profiles built from metadata (who you talk to, when, and how often)

This ecosystem — sometimes called surveillance capitalism — monetizes attention and behavior. Awareness and settings management help you stay in control.

Over-Dependence on AI

When your device writes your messages, chooses your routes, and remembers your tasks, it's easy to offload more than intended.

Potential downsides:

- reduced memory retention
- difficulty performing tasks without assistance
- erosion of problem-solving skills
- increased anxiety when the device isn't available

AI should support your thinking — not replace it.

Algorithmic Bias in Everyday Features

Many phone features rely on machine learning models. These can underperform for certain groups if the training data wasn't diverse enough.

Examples:

- facial recognition failing more often for darker skin tones
- predictive text that reinforces stereotypes
- voice assistants misinterpreting certain accents

Bias isn't just a societal issue — it can appear in your daily tools, too.

Privacy Tip

Review app permissions regularly — especially for:

- camera
- microphone
- location
- health data
- background activity

Disable any access that isn't essential. Small tweaks significantly reduce data exposure.

---

7. The Future of Mobile AI

Smartphones aren't just getting faster — they're becoming more perceptive, more autonomous, and more capable of collaborating with you. Many features that once felt futuristic already exist in early forms today. The next wave of mobile AI won't just add new tools — it will make the tools we already use far more powerful, seamless, and integrated.

On-Device AI Copilots

Local AI models are already appearing on premium phones, but they're about to expand dramatically.

What's evolving:

- faster, richer generative assistance that runs fully offline
- deeper integration across system apps (email, camera, notes, messages)
- real-time editing, summarizing, and reasoning without sending data to the cloud
- improved privacy because processing stays on your device

Future on-device copilots will feel less like an app and more like a system-level intelligence supporting everything you do.

Augmented Reality, But Matured

AR features already exist — from room scanning to plant identification — but they're early-stage.

What's next:

- more accurate scene understanding
- richer overlays (instructions, translations, labels) that adapt instantly as you move
- AR in everyday utilities, not just niche apps: shopping, maintenance, navigation
- seamless integration with glasses or heads-up displays, with the phone acting as the AI hub

The shift is from "AR that works sometimes" to "AR that you rely on."

Real-Time Translation, Beyond Words

Real-time translation already works well in earbuds and apps, but future versions will be far richer:

- smoother, more natural conversational flow
- better cultural and idiomatic awareness
- mixed-media translation (signs, menus, audio, handwriting)
- cross-lingual group conversations
- offline translation powered by on-device models

The experience will feel less like technology — more like speaking through a bilingual companion.

Health Diagnostics, Moving From Tracking to Understanding

Wearables already detect heart rate, sleep, and even irregular rhythms. Future diagnostics build on these foundations:

- early detection of conditions through subtle patterns in voice, gait, breathing, or typing
- context-aware health suggestions (stress-aware interfaces, rest reminders, sleep coaching)
- AI interpretation of long-term wellness trends, not just daily stats
- better integration between medical apps, devices, and healthcare providers

Today's wearables "measure." Tomorrow's will "interpret."

Emotion Recognition and Adaptive Interfaces

Phones already detect things like focus or inactivity. We're moving toward interfaces that adapt more intelligently:

- recognizing frustration, confusion, or urgency from tone or behavior
- adjusting suggestions, tone, or layout based on emotional context
- offering mindful nudges (breaks, breathing, movement)
- dynamically reducing notifications when stress indicators rise

Not a robot psychiatrist — but technology that's more responsive to how you feel.

Phones Become Partners, Not Tools

The direction is unmistakable:

Your phone is shifting from a device you operate to a digital partner that collaborates with you.

It will increasingly:

- anticipate needs
- adapt to routines

- help you interpret your environment
- simplify complex tasks
- reduce friction in daily life

The future of mobile AI isn't about replacing you — it's about amplifying you.

Quick Takeaways: AI in Your Pocket

» Your smartphone is an AI hub

AI powers core features you use daily — from voice assistants and camera intelligence to predictive typing, security, and smart battery management.

» Apps are smarter than ever

Navigation, photos, streaming, shopping, translation, and health tracking rely heavily on AI to personalize and improve your experience.

» On-device vs. cloud AI matters

Local processing offers speed and privacy; cloud processing delivers complex, large-scale intelligence. Knowing the difference helps you manage your data wisely.

» You can personalize far more than you think

Voice routines, predictive keyboards, accessibility tools, smart call features, and photo search can all be customized to fit your habits.

» Privacy is a choice — but only if you set it

Review permissions, limit background access, and use local-processing options where available.

» The future of mobile AI is already taking shape

On-device copilots, advanced AR, richer real-time translation, proactive health diagnostics, and emotion-aware interfaces are evolving quickly — turning phones into intelligent partners rather than passive tools.

Chapter Summary: AI in Your Pocket

Smartphones and wearables have quietly become powerful AI platforms woven into everyday life. They listen, see, predict, translate, classify, and personalize — often without users realizing just how much intelligence is happening behind the scenes.

AI enhances essential phone functions: voice assistants understand natural language, cameras optimize every shot, predictive keyboards learn your writing style, and on-device models keep interactions private and instantaneous. Third-party apps—from navigation and streaming to shopping and language learning—use machine learning to adapt to your preferences, anticipate your needs, and streamline daily routines.

Behind it all, your device constantly balances **on-device** and **cloud-based AI**, choosing the right approach depending on speed, privacy, and processing needs. With a few intentional adjustments — customizing routines, reviewing permissions, and enabling local features — you can make mobile AI more helpful and more aligned with your values.

Wearables extend this intelligence even further, offering real-time health monitoring, proactive wellness insights, and hands-free interactions. But these conveniences come with trade-offs: over-personalization, potential data misuse, algorithmic bias, and over-dependence.

Understanding how mobile AI works helps you use it wisely and protect your digital autonomy. The next wave — on-device AI copilots, mature AR, richer translations, health diagnostics, and emotion-aware interfaces — points toward a future where your phone becomes less of a tool and more of a collaborative assistant.

AI isn't just in the cloud — it's in your pocket, on your wrist, and integrated into your daily life. Used intentionally, it can save time, reduce friction, expand your capabilities, and help you navigate an increasingly complex world.

Chapter 16

CHATBOTS BEFORE CHATGPT — HOW EARLY AI SYSTEMS HELPED BUSINESSES COMMUNICATE

Introduction: Conversations Without People

Before generative AI tools like ChatGPT became part of everyday life, most people first interacted with artificial intelligence through a much simpler system: the traditional chatbot. These digital assistants — often found on websites, customer service portals, or automated phone menus — offered businesses a practical way to answer questions, automate transactions, and provide support around the clock.

But traditional chatbots worked very differently from the fluid, conversational systems we now take for granted. They didn't "understand" language in any meaningful sense. Instead, they followed scripts, recognized keywords, and moved users through predetermined paths. They were efficient and predictable — as long as the user stayed within their boundaries.

In this chapter, we'll explore what those early chatbots were, how they worked, where they showed up, and why businesses relied on them. Most importantly, we'll examine how fundamentally different they were from modern AI assistants powered by large language models (LLMs) — and why the leap from rule-based systems to generative models represents one of the biggest transformations in the history of conversational technology.

Understanding these earlier systems sets the stage for the next chapter, where we explore how tools like ChatGPT reshaped expectations, capabilities, and the very definition of what a "chatbot" can be.

---

1. What Is a Traditional Chatbot?

Before the rise of large language models, the word chatbot referred to a very different kind of system — one that was helpful, predictable, and strictly limited. A traditional chatbot is a rule-based software program designed to simulate conversation through structured, predefined interactions. Instead of "understanding" language, these bots operate within a controlled script, using decision trees, keyword triggers, or menu-like flows to guide the user.

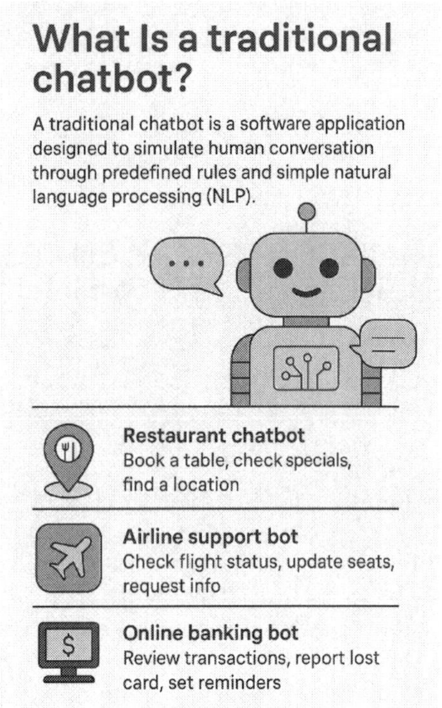

The easiest way to think about them is as digital assistants with a script. They perform well when tasks are repetitive and clearly defined, such as providing store hours, retrieving an account balance, or helping users reset a password. Because the conversation path is predetermined, the bot's job is not to interpret nuance — it's to recognize a narrow set of inputs and deliver the corresponding response.

Traditional chatbots rely heavily on "if-then" logic and limited natural language processing. If a user types "What are your hours?" the bot may recognize the keyword hours and retrieve the corresponding answer. But if the user phrases it slightly differently — "When do you guys open up again?" — the bot may fail completely, stall, or redirect the user to a human agent. Flexibility was never their goal; automation was.

These early systems powered a wide range of everyday interactions:

- **A restaurant chatbot** offering today's specials, reservation options, or operating hours — but breaking when the user types something unexpected.
- **An airline's support bot** that handles seating changes or gate information, but cannot interpret a follow-up question unless it matches the script.
- **An online banking assistant** that lets users check balances or report fraud, but quickly escalates anything nuanced to a human representative.

Businesses adopted traditional chatbots because they were fast, consistent, and always available. They could handle thousands of routine questions without human intervention, reducing costs and improving response times. For users, this meant immediate answers to common needs — no waiting in phone queues or navigating complex websites.

But their limitations were equally clear. Traditional chatbots couldn't adapt, generalize, or learn. They had no memory of past interactions, no understanding of context, and no ability to handle conversation outside their narrow domain. They didn't improve over time unless a developer rewrote the rules.

They were efficient tools — not conversational partners.

Understanding how constrained these systems were helps explain why the shift to LLM-powered assistants feels revolutionary. The difference isn't subtle; it's a complete redefinition of what machines can do with language.

---

2. How Traditional Chatbots Work

Behind their simple interfaces, traditional chatbots rely on a predictable, modular architecture. While modern conversational AI uses deep learning and massive neural networks, earlier chatbot systems depend on carefully engineered rules, structured flows, and tightly controlled logic. Their strength comes not from intelligence, but from design.

At a high level, the workflow looks something like this:

1. The User Interface

Every interaction begins with an input channel — a website widget, a mobile app, a messaging platform, or an automated phone system. These interfaces collect user messages (or voice commands converted to text) and pass them into the chatbot engine.

2. Intent Recognition

Once the text arrives, the bot attempts to determine what the user wants.

Traditional systems do this with:

- Keyword matching ("hours," "refund," "track order")
- Pattern recognition using regular expressions
- Simple NLP models that classify phrases into a small set of possible intents

This is not understanding. It's matching. If the expected phrase appears, the chatbot proceeds. If not, the bot asks the user to rephrase — or fails completely.

3. Dialogue Management

This is the heart of the chatbot. The dialog manager decides the next step:

- Select a scripted response
- Ask a follow-up question (e.g., "What is your order number?")
- Navigate to a different branch of a decision tree

- Escalate to a human agent
- Call an external tool or API

Traditional chatbot flows resemble **choose-your-own-adventure** paths — rigid but efficient when well-designed.

4. Backend Integrations

To provide useful information, the bot often connects to internal systems:

- A retail chatbot may query inventory or shipping databases
- A banking bot accesses account balances or transaction histories
- A travel bot pulls flight status or boarding times

These integrations make the bot seem "smart," even though it's simply fetching data through APIs.

5. Response Generation

Finally, the chatbot sends a reply — usually drawn from a fixed library of responses. Some systems had light customization (inserting a name or order number), but these replies were ultimately pre-written and static.

Why This Architecture Worked

Traditional chatbots excelled at tasks with:

- Predictable user questions
- Clear steps and defined outcomes
- High volume and low complexity

For businesses, they were a breakthrough: reliable, scalable automation for routine interactions.

But because every part of the system depended on predefined logic, these chatbots were fundamentally limited. They couldn't adapt to new wording, infer ambiguous intent, or handle conversations that wandered outside their rigid flows.

Traditional bots didn't learn — they followed instructions.

---

3. Where Traditional Chatbots Show Up

You've probably interacted with many of these bots already. They appear wherever businesses need fast, repetitive communication:

Customer Service:

Handling common questions like tracking an order, resetting a password, or explaining return policies.

E-Commerce:

Helping users browse products, check availability, apply promo codes, or complete checkout.

Healthcare:

Supporting routine tasks such as appointment scheduling, reminders, or simple pre-visit questions.

Education:

Automating enrollment guidance, answering admissions queries, or pointing students to campus resources.

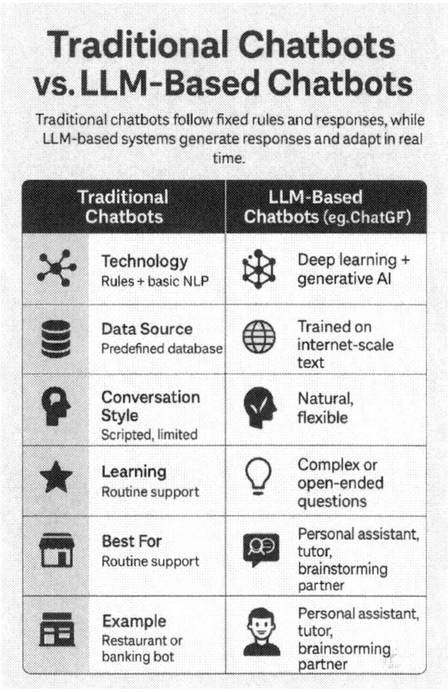

Finance:

Assisting with balance checks, recent charges, fraud alerts, or locating nearby ATMs.

In all of these domains, traditional chatbots shine because the conversations are predictable, structured, and high-volume.

---

4. Why Businesses Use Chatbots

Traditional chatbots became widespread because they delivered immediate, practical benefits — especially for organizations handling large volumes of routine questions.

Time-Saving:

They automate repetitive inquiries, freeing human staff to focus on complex or sensitive issues.

Always Available:

Bots offer 24/7 support, which is especially valuable for global customers or after-hours requests.

Cost-Effective:

They reduce call center load and lower operational costs by deflecting common queries.

Fast to Deploy:

Many platforms allow businesses to configure simple bots with minimal technical skills.

Customizable:

Companies can tailor scripts, connect bots to databases, and automate workflows like order tracking or appointment scheduling.

For small and mid-sized businesses, these advantages meant chatbots could scale customer support without expanding headcount.

---

5. Limitations of Traditional Chatbots

While traditional chatbots were useful, their limitations were easy to spot.

Rigid Conversations:

They followed scripts. If a user went off-path, used unusual phrasing, or made a typo, the bot often returned irrelevant answers or reached a dead end.

No Memory:

They couldn't recall prior interactions unless explicitly programmed. Each new question felt like starting from zero.

No Real Understanding:

Because they relied on keywords and rules—not context or reasoning—they struggled with nuance, emotion, or follow-up questions.

Narrow Scope:

They were designed for a single domain or a specific workflow. They couldn't generalize or transfer knowledge to new tasks.

Not Adaptive:

Traditional bots didn't learn from user interactions. If thousands of people hit the same error or asked for the same feature, the system would never improve unless a developer manually updated it.

This rigidity made them effective for predictable, repetitive tasks, but poorly suited for open-ended conversation, creative work, or complex reasoning — the kinds of things modern AI assistants now excel at.

These limitations are also the reason modern customizable AI assistants exist today — including the types now built within the Mastering AI ecosystem — which are designed to be more flexible, context-aware, and adaptive than traditional scripted bots ever could be.

6. Comparing Traditional Chatbots to Generative AI (LLMs)

To understand how dramatically conversational AI has evolved, it helps to compare traditional chatbots with today's large language model (LLM) systems like ChatGPT, Gemini, and Claude.

Traditional chatbots follow fixed rules. They rely on predefined responses and simple intent detection, making them reliable when interactions are predictable and tightly scoped. If you ask, "Where's my order?" they search a database and return a scripted answer. But if a question falls outside their flowchart, they stall, redirect, or fail.

Generative AI works very differently. LLMs are powered by deep learning models trained on massive datasets — books, articles, websites, code, transcripts. Instead of depending on scripts, they interpret meaning in context, adapt to user intent, and generate new text on the fly. They can summarize documents, brainstorm ideas, write emails, answer follow-up questions, and shift tone or format based on your request.

Where traditional chatbots automate **tasks**, LLMs support **conversation**.

Where traditional bots handle **known questions**, LLMs respond to **open-ended ones**.

Where traditional bots follow **rules**, LLMs rely on **reasoning patterns** learned during training.

This shift transforms what's possible. Businesses can move from menu-based bots to assistants that explain policies, help draft messages, guide troubleshooting, or offer personalized suggestions. LLMs don't replace task-specific bots — but they redefine what "chat" can mean in everyday tools.

---

7. When Traditional Chatbots Still Make Sense

Even with the rise of generative AI, traditional chatbots haven't disappeared — and in many cases, they remain the smarter, more reliable choice. Their simplicity is an advantage when businesses need predictable, fast, and controlled interactions.

Traditional chatbots are still ideal when:

The task is repetitive and well-defined.

Questions like "Where's my package?" or "Reset my password" don't require creative reasoning — they just require speed.

Privacy or compliance limits data sharing.

Some industries (like healthcare or finance) need strict control over what information is processed and how. A rule-based bot is easier to certify and audit.

Consistency matters more than flexibility.

If every customer must receive the same answer or follow the same steps, a scripted bot ensures uniformity.

User expectations are narrow.

When someone opens a support chat specifically to get a delivery update or check a reservation, a simple bot is efficient and appropriate.

Business workflows depend on precise logic.

For logistics, banking, or support automation, predictable branching logic often performs better than open-ended dialogue.

Put simply:

If the job doesn't require creativity, context understanding, or nuanced conversation, a traditional chatbot often remains the most efficient tool.

Generative AI expands what's possible — but the older style of bots still excels at doing one job well, every single time.

---

8. Looking Ahead

Traditional chatbots solved a specific problem: they automated predictable conversations. But the next generation of chat systems — powered by large language models — represents a shift from scripted interaction to true conversational intelligence.

Where early bots followed rigid flows, modern AI assistants can:

- Understand open-ended language
- Ask clarifying questions
- Adjust tone and style
- Handle complex, multi-step tasks
- Learn from context within a conversation

This evolution marks more than an incremental upgrade. It's a transition from menu-based automation to context-aware collaboration.

In the next chapter, we'll explore these new AI assistants in depth — how they generate responses, adapt to different users, and support everything from writing and analysis to creative work and task automation. Most importantly, we'll look at why this leap from rules to reasoning is reshaping how people and businesses communicate.

Quick Takeaways: Chatbots Before ChatGPT

» Rule-Based Systems

Traditional chatbots follow predefined scripts, decision trees, and basic NLP — ideal for predictable, repetitive tasks.

» Efficiency Over Flexibility

They excel at high-volume, low-complexity questions such as order tracking, appointment scheduling, or account lookups.

» Limited Understanding

These bots can't improvise, handle nuanced conversation, or adapt to unexpected phrasing. They respond only within their programmed boundaries.

» Still Useful Today

In settings where speed, control, and consistency matter — or where privacy and compliance restrict data sharing — traditional chatbots remain the right tool.

» A Foundation for What Comes Next

Understanding how these systems work sets the stage for appreciating the leap to LLM-powered assistants, which engage in open, creative, and dynamic dialogue.

Chapter Summary

Before generative AI became mainstream, traditional chatbots were already automating routine conversations across customer service, retail, finance, healthcare, and education. Built on rule-based logic and simple natural language recognition, these systems offered businesses a cost-effective way to handle predictable, high-volume inquiries around the clock.

Their strength was reliability — not creativity. They followed scripts, couldn't adapt to unusual phrasing, and lacked memory or contextual understanding. Despite these limits, they played a crucial role in improving response times, reducing support workloads, and providing consistent experiences where users needed structured information fast.

While today's AI tools can analyze context, generate original responses, and assist with complex tasks, traditional chatbots still have a place in environments where predictability, compliance, and tight control are essential. They remain part of the hidden infrastructure that keeps many digital interactions running smoothly.

In the next chapter, we'll explore how large language models transformed this landscape — turning simple automated responders into conversational partners capable of reasoning, writing, and collaborating in ways early chatbots never could.

Chapter 17

TALKING TO AI — HOW TO USE CHATGPT, GEMINI, AND OTHER TOOLS EFFECTIVELY

Introduction: A New Kind of Assistant

In the past, chatbots followed strict scripts. They could help you track a package or confirm store hours, but ask anything unexpected and they'd fall flat. That changed with the arrival of **large language models (LLMs)** — a new generation of AI systems capable of understanding and generating human language in flexible ways. Tools like ChatGPT, Google Gemini, and Claude introduced a new era of adaptability, creativity, and contextual understanding.

Suddenly, chatbots weren't just answering questions — they were **writing stories, tutoring students, summarizing documents, generating code, crafting emails, helping with business planning**, and much more.

This chapter explores what makes LLM-powered chatbots different, how they work, what they can (and can't) do, and how to get the most out of them. Because when it comes to getting helpful results, it's not just about the tool you choose — it's about how you use it.

---

1. What Are These Tools, Really?

At the core of ChatGPT, Google Gemini, Claude, and similar tools is something called a **large language model (LLM)**. These are advanced AI systems trained on enormous collections of human-written text — books, websites, code, conversation transcripts, and more. The goal is to learn the structure, style, and relationships within language so well that the model can generate text that feels natural, informative, and often surprisingly creative.

But it's just as important to understand what they are not.

LLMs don't think or reason like humans. They don't "know" facts in the way a person does, and they don't have beliefs, goals, or self-awareness. Instead, they work by **predicting the most likely next word (or token)** in a sentence, based on your input and everything they learned during training.

That might sound mechanical — and it is — but when you combine this mechanism with billions (or even trillions) of **parameters** and a long context window, you get something that feels conversational, knowledgeable, and adaptable.

How They Differ from Traditional Software

Traditional software follows fixed instructions:

if X happens, do Y. Every possible situation has to be anticipated in advance.

LLMs behave differently:

- They don't follow rigid, preset rules — they adapt in real time to how you ask a question.
- They don't just retrieve an answer — they generate new text from scratch.
- They can change tone, style, and behavior depending on what you ask, how you ask, and which "role" you assign.

For example, you can say:

- "Act like a motivational coach,"
- "Explain this to me like I'm 12," or
- "Write this in a formal, corporate tone."

And the model will adjust accordingly.

You can also ask it to summarize a 20-page PDF, review some code, or help brainstorm ideas for a marketing campaign. This flexibility is part of what makes LLM-based tools feel so humanlike — and so powerful.

Under the surface, though, they're still doing one core thing very well: using patterns learned from huge amounts of text to predict what comes next. The magic is not that they "understand" in a human sense, but that their predictions are good enough to be genuinely useful.

---

2. Comparing the Most Popular LLM Chatbots

Although ChatGPT, Gemini, Claude, and Pi all use large language models, each one has a distinct style and set of strengths. Choosing the right one depends on what you need: creativity, precision, long-form reasoning, or a more conversational tone.

ChatGPT

Best for: structured thinking, coding, analysis, versatile writing

What stands out: adapts well to roles ("act as a tutor…"), excellent at breaking down tasks, strong at step-by-step reasoning

Ideal for: planning, drafting, creating content, learning technical topics

Google Gemini

Best for: search-connected answers, real-time information

What stands out: integrates smoothly with Google tools, strong at retrieving up-to-date info

Ideal for: fact-checking, researching, quick summaries, everyday questions

Claude

Best for: long, thoughtful, calm explanations

What stands out: excellent at extended reasoning, document analysis, and maintaining coherent tone across long inputs

Ideal for: studying, reviewing PDFs, reflective writing, analysis

Pi (Inflection AI)

Best for: friendly, supportive conversation

What stands out: emotional awareness, gentle style, simple guidance

Ideal for: reflective chats, motivation, coaching-like conversations

Free vs. Paid Access

Most tools offer both free and subscription-based tiers. In general:

Free plans:

- Access to smaller models
- Basic chat and Q&A
- Limited memory and slower performance

Paid plans:

- Advanced models
- Longer context windows
- File and image uploads
- Faster, more reliable responses

Is upgrading worth it?

If you use these tools for writing, research, studying, business, or creativity — yes. If your use is casual or occasional, the free versions already offer strong capabilities.

3. How LLM-Based Chatbots Work

It's easy to think of tools like ChatGPT or Gemini as "smart apps," but they operate very differently from traditional software. Instead of following fixed instructions, they rely on large language models (LLMs) — neural networks trained on vast amounts of text to understand and generate human-like language. Here's how that process works behind the scenes.

1. Tokenization: Turning Language Into Pieces the Model Can Read

Before an LLM can interpret a message, it breaks your text into tokens — small parts such as whole words, subwords, or even characters.

For example, the prompt:

"Tell me a story about a dragon who loves pizza."

gets split into individual tokens. These are then converted into numerical representations the model can process.

2. Context Modeling: Understanding Meaning Through Surrounding Text

LLMs don't look at each token in isolation. They analyze your entire message — and the conversation so far — to interpret tone, references, and intent.

This lets them:

- follow the flow of a dialogue
- keep track of earlier details
- resolve ambiguous phrases ("Can you move that?")
- maintain consistent style or voice

This context-awareness is what makes conversations feel coherent across multiple exchanges.

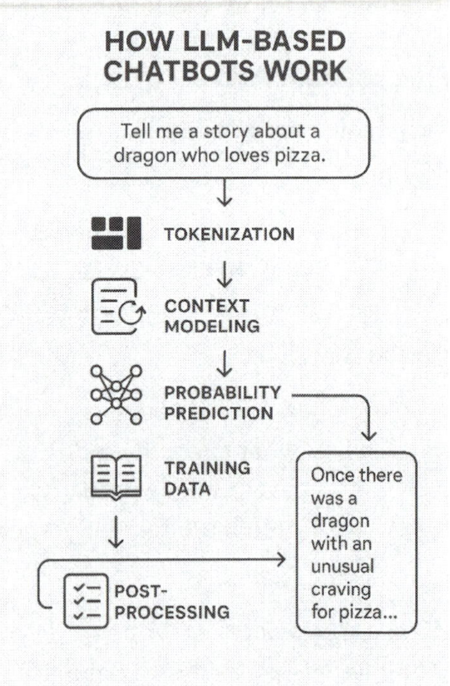

3. Probability Prediction: Generating the Next Word, One Step at a Time

At the core, LLMs don't "know" answers — they predict them.

They generate text one token at a time by calculating which token is most likely to follow next, based on patterns learned during training.

When you ask:

"Write a poem about a lighthouse in a storm,"

the model isn't retrieving a stored poem. It's generating one on the spot by predicting each token in sequence.

4. Training Data: Where the Model Learns Its Patterns

During training, the model is exposed to billions of sentences from books, articles, websites, manuals, dialogues, and more.

It doesn't memorize everything — it learns patterns such as:

- how questions and answers typically connect
- how stories, emails, or explanations are structured
- relationships between concepts (e.g., mountains → hiking → elevation)

This pattern learning allows the model to produce text that feels meaningful and structured even though it isn't "thinking" in a human sense.

5. Personalization and Fine-Tuning: Adapting to Your Needs

Some chatbots adapt based on:

- previous messages within a session
- the tone you prefer (casual, formal, humorous)
- the role you assign ("Act like a tutor," "Act like a marketer")
- uploaded files or images you provide

Some platforms also allow longer-term memory, enabling the assistant to recall your preferences across conversations.

6. Post-Processing: Formatting and Guardrails

After generating a draft response, many systems apply additional layers:

- formatting tools (lists, tables, markdown)
- relevance checks
- filters to avoid unsafe or inappropriate content

This is why structured answers like summaries, outlines, or step-by-step guides appear so cleanly formatted.

Why This Matters

Understanding this process helps you use these tools more effectively.
It explains:

- why they can be remarkably creative
- why they sometimes make mistakes or "hallucinate"
- why clear prompting improves results
- why they need human direction and oversight

LLMs are powerful, but they are not autonomous thinkers — they're sophisticated pattern generators that rely on your clarity to produce their best work.

---

4. Getting the Most from a Chatbot: How to Ask Better Questions

LLM-based chatbots respond best when given clear direction. Think of prompting as guiding a very capable — but very literal — assistant. The clearer your request, the better the result.

Here are ways to sharpen your prompts and get more consistent, useful responses:

Be Specific

Replace vague instructions with clearly defined tasks.
Instead of:

"Write something about health,"

try:

"Write a 300-word blog post with three simple heart-health tips for people over 50."

Use Roles

You can assign a perspective or expertise:

- *"Act like a career coach."*
- *"Explain this as a high school science teacher."*
- *"You're a marketing strategist—rewrite this pitch."*

Roles instantly shift tone, structure, and focus.

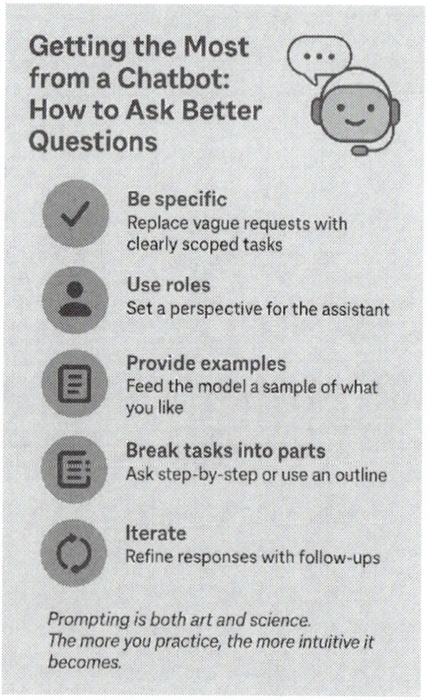

Provide Examples

If you have a preferred style or format, show it.

- *"Here's how I wrote it. Improve it, but keep the tone conversational."*

LLMs learn quickly from patterns you supply.

Break Tasks Into Parts

Complex requests work better when approached step-by-step:

- *"First, create an outline."*
- *"Now expand section 2."*
- *"Summarize this in 150 words."*

Iterate and Refine

Follow-up prompts are where the real power lies:

- *"Make this more concise."*
- *"Add a friendlier tone."*
- *"Shorten this but keep the key points."*

A single prompt is rarely the final product — good prompting is conversational.

Real-World Tip

You can combine prompting with voice input on mobile, or use file uploads and multimodal features to analyze documents, images, or transcripts.

If you want to go deeper, the Mastering AI website also includes short practical guides on asking better questions — useful for refining your prompting skills as you explore these tools.

---

5. Everyday Uses — What These Tools Can Do for You

LLM-based chatbots aren't just for tech experts. They're versatile tools that can support everyday tasks across writing, planning, learning, shopping, and productivity. With the right prompts, they function like flexible personal assistants.

Writing

- Draft or refine emails, reports, resumes, and blog posts
- Rewrite text for clarity or adjust tone (formal, friendly, concise)
- Create summaries of long documents or transcripts

Planning

- Build trip itineraries, packing lists, or budget breakdowns
- Plan meals, events, or weekly schedules
- Organize tasks into lists, timelines, or workflows

Learning

- Explain complex topics in simple language
- Generate quizzes or study guides
- Summarize readings, research papers, or chapters

Shopping & Decision-Making

- Compare product features or prices
- Suggest gifts based on interests, age, or budget
- Interpret reviews and highlight pros / cons

Productivity

- Clean up messy notes or meeting transcripts
- Create templates for projects, checklists, or processes
- Extract key takeaways from long text

These tools adapt to your input, so the more context you provide, the more tailored the results become.

Industry Crossover

While this chapter focuses on personal use, LLMs also support:

- **Healthcare:** appointment reminders, basic guidance (not diagnosis)
- **Education:** tutoring, practice questions, language learning
- **Retail:** product recommendations, customer support
- **Workplace tools:** summarization, research, and internal support assistants

Used thoughtfully, these capabilities help simplify daily life and save time — whether you're drafting a message, planning a trip, or learning something new.

---

6. Business and Professional Use Cases

LLMs are becoming indispensable in workplaces of all sizes. Whether you're a solopreneur, part of a small team, or working inside a large organization, AI can streamline everyday tasks and spark new ideas — without replacing your judgment or expertise.

Marketing

- Draft ads, landing pages, newsletters, or social media captions
- Generate content ideas, keyword lists, and SEO-friendly blog outlines
- Adapt messaging for different audiences or platforms

Operations & Productivity

- Summarize long documents, meeting notes, or email threads
- Draft internal communication or project updates
- Create templates for reports, workflows, and procedures

Software & Engineering

- Generate or explain code snippets
- Suggest improvements, debug errors, or document functions
- Translate technical concepts into plain language for teammates or clients

Sales & Strategy

- Write outreach emails, proposals, and pitch drafts
- Brainstorm product features, pricing models, or market segments
- Prep for meetings with summaries and talking points

Research & Analysis

- Break down academic papers, industry reports, or data summaries
- Extract insights across multiple documents
- Organize findings into bullet points, tables, or concise briefs

The Trend Behind the Scenes

Many companies now embed LLMs into internal systems — HR support bots, IT assistants, finance helpers, onboarding guides — to reduce repetitive tasks and improve response times.

These tools don't replace teams. They accelerate them.

Why This Matters

LLMs act like tireless drafting partners:

- They never get stuck on a blank page
- They can produce variations instantly
- They help professionals focus on higher-level thinking rather than repetitive work

Even small businesses benefit — using AI to draft content, manage communication, and organize information without hiring additional staff.

---

7. How to Prompt Like a Pro (Prompt Engineering Lite)

Prompting is a communication skill. The clearer and more structured your instructions, the better an LLM performs. You don't need advanced "prompt engineering" knowledge — you just need to guide the AI the same way you'd guide a smart assistant who wants clear direction.

Here's how to get consistently better results:

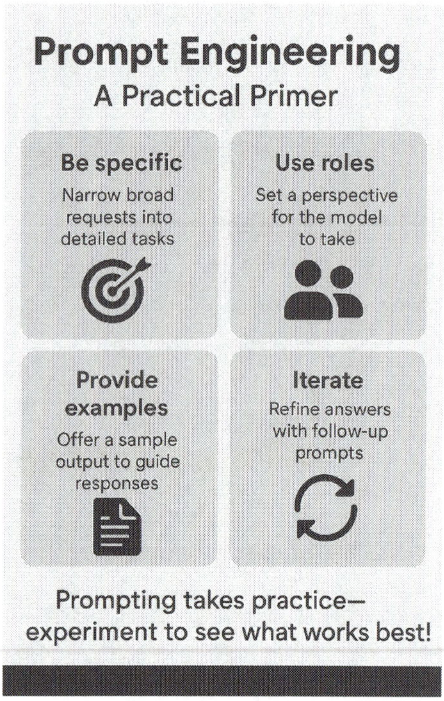

Role-Based Prompting

Give the model a perspective or expertise to adopt.

- *"You are a social media strategist. Help me create a week of posts."*
- *"Act as a financial advisor and explain the pros and cons of each option in plain language."*

This helps the model understand your expectations immediately.

Step-by-Step Instructions

Break work into phases so the AI doesn't jump ahead.

- *"First give me an outline. After I approve, expand each section."*
- *"List the pros and cons first. Then write a short recommendation."*

This improves structure and reduces rambling.

Feedback Loops

Treat the output as a draft, not a final product.

- "Make this more professional."
- "Shorten by 40% but keep the tone friendly."
- "Rewrite this using simpler language suitable for teens."

Refining in iterations leads to better results than trying to get it perfect in one prompt.

Format Control

Tell the model how you want the answer delivered.

- *"Give me this in a clean bullet list."*
- *"Output as a table with columns: Step, Description, Example."*
- *"Summarize this in 120 words."*

Good formatting makes the result more usable immediately.

Tone Control

Guide the voice of the content just as you would with a human editor.

- *"Use a warm, empathetic tone."*
- *"Make it funny and conversational."*
- *"Keep it concise and analytical."*

Use a professional, conversational tone

Use Tools When Available

Many LLM platforms allow file uploads, image inputs, or integrated tools like code execution, document analysis, or browsing.

Using them can dramatically expand what a prompt can achieve.

The Real Secret

Clear intention + structure + examples = consistently excellent results.

You don't need long prompts.

You need **purposeful prompts.**

Prompting isn't a trick — it's collaboration.

---

8. Things to Watch Out For

Even though these tools are powerful, they're not perfect. Using them wisely means keeping a few limitations in mind — not to avoid them, but to guide them with intention.

Hallucinations

LLMs can sometimes generate information that sounds confident but isn't correct.

They rely on patterns in language, not factual certainty, so always double-check important claims — especially for work, health, legal, or financial topics.

Bias in Training Data

Because models learn from human-created text, they may reflect social, cultural, or historical biases.

Be cautious when using them for decisions that affect people, especially in sensitive contexts. Human judgment should always remain the final filter.

Privacy and Sensitive Information

Avoid sharing personal identifiers, financial information, or confidential business data.

LLMs are excellent for drafts, ideas, and explanations — not for storing or transmitting private data.

Over-Reliance

These tools can speed up tasks dramatically, but they're not experts.

Use them to generate options, not to replace your own thinking or professional review.

Knowledge Gaps

Some models have fixed training windows or cannot access up-to-date information unless browsing is explicitly enabled.

If your question requires real-time information, verify through a trusted source.

Why This Matters

These limitations don't diminish the usefulness of LLMs — they strengthen it.

The better you understand the boundaries, the more effectively you can direct the tool, validate its outputs, and treat it as a collaborator rather than an authority.

---

9. Wrapping Up: It's Not Magic — It's Collaboration

Large language models feel powerful, but they're not mysterious. They excel when you provide direction, and they struggle when the goal is vague. The best way to think about them is as highly capable collaborators — assistants that can draft, refine, explore, and generate, but still depend on you for judgment, context, and goals.

They don't replace your thinking; they amplify it.

When you bring clarity, they bring acceleration.

When you bring structure, they bring creativity.

When you bring the goal, they help build the path.

But like any collaborator, they need guidance, correction, and boundaries. They won't always get things right, and they don't "understand" the way humans do. They predict patterns — and they rely on you to steer the conversation toward something useful.

Used well, LLMs can:

- Help you think through problems
- Speed up tedious or repetitive work
- Improve your writing and communication
- Expand your creative output
- Break complex tasks into simple steps
- Offer ideas you may not have considered

Human + AI is stronger than either alone.

The future is not about replacing people — it's about working alongside tools that enhance what you already do well.

The more intentional you are, the more these systems become reliable partners: fast, versatile, and tireless — but ultimately shaped by your direction.

---

10. Practical Example:
A Real Conversation with a Travel Planner AI Assistant

To bring everything together, let's walk through a real-world scenario where a user interacts with an LLM-based assistant. This example will show how effective prompting, personalization, and dialogue flow can result in useful and specific responses.

Scenario: Planning a Vacation with an AI Travel Assistant

1. **Setup**: Name the chat "Travel AI Assistant." Use any LLM-based tool like ChatGPT, Gemini, or Claude.

2. **Prompt Instructions**: Set the assistant's role with the following instruction:

   Imagine you are a professional travel advisor named WanderMate. You help users plan vacations, find destinations, create itineraries, and offer personalized travel tips based on their preferences, budget, and travel goals.

3. **Initial Prompt**: Now, as the user, begin the chat with the following message:

   Hi WanderMate, I want to plan a 7-day vacation to Europe this summer. I'm on a moderate budget, enjoy history and local food, and prefer walkable cities. Can you help me get started?

Sample Output:

Absolutely! I'd love to help you plan your trip. Based on your interests and budget, here's a suggested itinerary for a 7-day vacation:

Day 1-3: Lisbon, Portugal

- Affordable and charming city with great food and walkable neighborhoods
- Explore the Alfama district, take a tram ride, and enjoy traditional Fado music
- Visit historic sites like Jerónimos Monastery and Belém Tower

Day 4-7: Prague, Czech Republic

- A budget-friendly destination full of medieval charm
- Discover Prague Castle, Charles Bridge, and the Astronomical Clock
- Try local dishes like svíčková and trdelník in Old Town Square

Estimated Budget: €1,200–€1,500 (excluding flights)

Travel Tips:

- Use budget airlines for intra-European flights (e.g., Ryanair, EasyJet)
- Stay in boutique hostels or guesthouses with high ratings
- Consider getting a local SIM card for easier navigation

Would you like help booking accommodations or adjusting the trip to fit specific dates?

4. **Follow-up Prompt**:

That sounds great! I'm traveling in July. I'd prefer to avoid large tourist crowds and would love to include a small coastal town if possible. Can we swap one city for a seaside destination?

Sample Output:

Of course! July can get busy, so here's a revised plan:

Day 1-3: Porto, Portugal

- A beautiful coastal city known for its port wine, riverside scenery, and relaxed vibe
- Visit Ribeira, Livraria Lello, and take a boat cruise along the Douro River

Day 4-7: Ljubljana, Slovenia

- A peaceful, green capital city with easy access to Lake Bled for a scenic day trip
- Enjoy open-air cafes, the Ljubljana Castle, and riverside walks

Bonus Tip: Book early morning sightseeing tours to beat the crowds and take advantage of cooler weather.

5. **Why This Matters**: This example shows how LLM chatbots can:
    - Interpret vague goals and convert them into structured, actionable suggestions
    - Adapt based on preferences like budget, interests, and time of year
    - Offer follow-up questions or suggestions to refine and personalize the experience

By combining context, specificity, and iterative dialogue, these assistants can deliver highly relevant support—much like a human travel agent.

You can apply this same process for other use cases—whether you're getting resume feedback, building a workout plan, or writing a business proposal. The key is in how you frame your request.

End Note: Try your own prompt now. Choose a role, define your goal, and see how much your chatbot can do for you!

The same prompting fundamentals you see here are also taught throughout Mastering AI's guides and videos, so you can apply them across any task or tool.

Quick Takeaways: Talking to AI Tools

» Text-Based Intelligence

AI assistants like ChatGPT, Gemini, Claude, and others are powered by large language models (LLMs) that understand and generate human-like text across thousands of tasks.

» Choose the Right Fit

Each tool has its own strengths — reasoning, creativity, long-form thinking, conversational tone, or real-time search — so match the assistant to the job.

» Prompting is a Skill

Clear, specific, and well-structured instructions consistently produce better results. Learning how to ask good questions is the fastest way to improve what you get back.

» Everyday Utility

LLMs can help with writing, planning, learning, research, brainstorming, coding, decision support, and task automation — no technical background required.

» Be Aware of Limitations

These systems can make mistakes, reflect bias, or handle data imperfectly. Verify important information and avoid sharing sensitive or confidential details.

» Human in the Loop

Think of AI as a collaborative partner — fast and capable, but still needing your direction, oversight, and judgment.

Chapter Summary

This chapter introduced large language model–based chatbots — tools like ChatGPT, Gemini, Claude, and Pi — and explained what makes them fundamentally different from traditional rule-based chatbots. Instead of following fixed scripts, these systems generate language dynamically, adapt to context, and handle a wide range of tasks, from writing and planning to analysis, education, and creative work.

We explored how LLMs work at a high level: by breaking language into tokens, interpreting context, and predicting the next word using patterns learned from enormous training datasets. We also covered the practical skills needed to get useful results, especially the importance of clear prompting, role-setting, and iterative refinement.

From personal use to business applications, these tools have become powerful everyday assistants — helping people write better, think faster, and explore ideas more deeply. But they still require human oversight. Issues like hallucinations, bias, outdated information, and privacy concerns make it essential to use them thoughtfully.

The key takeaway is that talking to AI is a collaboration. You bring the clarity, intent, and judgment; the AI brings speed, structure, and creativity. Together, they create new possibilities for productivity and problem-solving across every part of daily life and work.

Chapter 18

AI AT HOME
SMARTER LIVING THROUGH INTELLIGENT DEVICES

Introduction: The Connected Home

Home automation used to mean something simple — like setting a timer on your coffee maker. Today, thanks to AI, our living spaces can adjust themselves automatically, learn from our routines, and support our comfort, safety, and efficiency in ways that would have seemed futuristic only a decade ago.

But AI in the home isn't just about convenience. It represents a shift in how we interact with our environment. A modern smart home isn't just a collection of gadgets — it's an adaptive system that senses, interprets, and responds — often powered by AI models running quietly in the background.

This chapter explores what actually makes a home "smart," what AI is doing behind the scenes, how to personalize it responsibly, and how to balance convenience with privacy and control. The goal isn't to fill your home with devices — it's to help you understand how AI can enhance everyday living without taking over your autonomy.

---

1. What Makes a Home "Smart"?

A "smart home" isn't defined by how futuristic it looks, but by how intelligently and automatically it responds to your needs. At its core, a smart home is a living space equipped with interconnected devices — lighting, climate control, appliances, security systems, and entertainment — that are enhanced with artificial intelligence and connected to the internet. These devices don't just respond to commands; many of them learn from your behavior, adapt to your routines, and can act proactively on your behalf.

What separates a smart home from a home filled with gadgets is **intelligence and integration**. A truly smart home learns from experience. It knows when you typically wake up, what temperature you prefer in the evenings, when to lock the doors, and how to adjust lighting as the day fades. It recognizes patterns, anticipates your preferences, and can even optimize for energy savings or safety.

These behaviors come from predictive machine learning — not from generative AI — which means the system is recognizing patterns, not 'thinking' in a conversational sense.

Beyond Remote Control

In the early days, home automation meant controlling devices from your phone — turning lights on remotely or adjusting the thermostat on the way home. Today, smart homes go a step further: they **make decisions for you** based on AI-powered models. For example:

- A smart thermostat might lower the temperature when it detects you've left for work — even if you forgot.
- A smart speaker might suggest turning off the oven because you've started your "bedtime" routine.
- A security camera might distinguish between your neighbor's cat and a potential intruder — and alert you only when necessary.

Learning Routines and Preferences

The heart of smart home AI is **pattern recognition**. Devices observe your habits over time:

- What time do you wake up?
- Do you prefer bright or dim lighting in the evening?
- How often do you leave windows open?
- When do you usually leave and return home?

Using machine learning, the system builds a model of your behavior and preferences. It doesn't just react — it **predicts**. For example, your smart lights might automatically adjust as sunset approaches, or your blinds may lower if it's a hot afternoon and you typically close them to keep cool.

Sensors, Connectivity, and Control

Smart homes rely on a web of technologies working together:

- **Sensors** detect motion, temperature, humidity, sound, and light.
- **Connectivity** (Wi-Fi, Bluetooth, Zigbee, Z-Wave) enables communication between devices and central hubs.

- **Control interfaces** — such as smartphone apps, touch panels, or voice assistants — give you input and visibility, while AI models quietly handle the complexity.

In a well-integrated system, these components combine to create **ambient intelligence** — a home that feels responsive, not reactive. You don't need to press buttons constantly; instead, the environment aligns itself with your life.

The Role of AI: From Static to Adaptive

What makes the smart home more than just convenient is AI. Without it, automation is static — you create a rule, and the device follows it. With AI, automation becomes **adaptive**. The home starts to:

- Detect anomalies (e.g., "You're usually home by 6 p.m. — should I alert someone?")
- Offer proactive suggestions (e.g., "You've left the lights on in the garage again — should I auto-shut them off next time?")
- Adjust automatically without requiring your constant input

This shift — from control to collaboration — is what defines the modern smart home experience. It's not just about pressing buttons or issuing commands. It's about having a home that understands, assists, and evolves with you.

A Living System

In essence, a smart home is like a digital nervous system for your house. It constantly senses, interprets, and responds to its environment. Just like a human body adapts to heat, noise, or motion, a smart home can do the same — turning on fans, adjusting music, or alerting you to unexpected activity. Over time, the system becomes a kind of **living infrastructure**, quietly optimizing your daily life.

---

2. Common AI-Powered Home Devices

Many homes now include AI-driven tools that operate in the background:

- **Smart thermostats** (e.g., Nest, Ecobee): Learn your habits and adjust heating / cooling to optimize comfort and energy efficiency.
- **Smart lights** (e.g., Philips Hue): Use motion detection or routines to adjust lighting and reduce energy waste.
- **Voice assistants** (e.g., Alexa, Google Assistant): Control devices, answer questions, manage shopping lists, and more.
- **Security systems**: AI-enhanced cameras and doorbells (e.g., Ring, Arlo) recognize faces, detect unusual movement, and send real-time alerts.

- **Cleaning robots** (e.g., Roomba, Roborock): Use AI-based path planning to vacuum or mop efficiently, avoid obstacles, and return to base.

These devices often integrate with broader systems like Apple HomeKit or Samsung SmartThings.

---

3. What AI Actually Does Behind the Scenes

Most smart home tools don't use large-scale generative AI (like ChatGPT). Instead, they rely on more focused machine learning algorithms that:

- Recognize patterns in behavior (when you adjust the thermostat, turn lights on/off)
- Use classification (e.g., is that a person or a package on your porch?)
- Optimize energy or task efficiency based on feedback loops
- Send data to the cloud for broader updates or improvements

While it may feel like magic, AI is mostly doing predictive analysis based on past input and sensor data. Most of these models are relatively small and efficient, designed for single-purpose tasks like classification or motion detection.

---

4. Customizing Your Smart Home Intelligently

Smart home tools work best when you actively guide them with clear routines and settings. A few ways to do this:

- **Create routines**: "Goodnight" turns off lights, lowers thermostat, locks doors.
- **Use voice shortcuts**: Set up custom phrases that trigger multi-step actions.
- **Adjust schedules**: Most systems let you set specific times or conditions for devices to activate.
- **Review app dashboards**: Devices often offer usage history that helps you fine-tune performance.

Feature Box: Real-World Examples of Smart Home Automation

Use these examples to spark ideas for how you can automate parts of your home:

▸ **"Good Morning" Scene** (Control4, Apple Home)

- Lights slowly brighten
- Thermostat adjusts to wake temperature

- Coffee machine starts
- Weather + schedule read aloud

► **Smart Entry and Security**

- Doorbell cam notifies you when a known face arrives
- Smart locks unlock for trusted people via geofencing
- Outdoor lights turn on if motion is detected late at night

► **"Movie Night" Mode**

- Lights dim, blinds close
- TV switches to Netflix or your preferred app
- Surround sound adjusts automatically

► **Climate Optimization**

- Smart thermostat learns your patterns
- Runs fans when humidity is high
- Saves energy when nobody's home

► **"Bedtime" Routine**

- Turns off all lights
- Locks all doors
- Plays white noise or meditation audio

► **Voice Control Everywhere**

- "Alexa, turn off the kitchen lights."
- "Hey Siri, lock the front door."
- "OK Google, set the thermostat to 70."

► **Remote Control Anytime**

- Turn off devices or check cameras from anywhere
- Let in guests or deliveries while away from home

► *Start small—one room, one routine. Build from there as your comfort and needs grow.*

---

5. Risks and Responsible Use

While smart homes offer benefits, there are important trade-offs to consider:

- **Surveillance**: Always-on microphones and cameras may feel invasive, especially if not clearly disclosed.
- **Data privacy**: Devices may collect sensitive data (e.g., when you're home, how you sleep) that is stored in the cloud.

- **Third-party access**: Some manufacturers or apps may share or sell your data to advertisers depending on their privacy policies and regional regulations.
- **Security vulnerabilities**: Poorly secured smart devices can be exploited by hackers to gain access to your network.

Checklist: Regularly update firmware, use strong passwords, disable unnecessary features, and check app permissions.

---

6. Accessibility and Inclusive Design

AI at home isn't just about convenience—it can significantly improve independence for older adults or people with disabilities:

- Voice control replaces hard-to-reach switches
- Motion-activated lights prevent falls
- Smart sensors monitor for emergencies like fire, gas leaks, or water damage
- AI cameras can detect if someone has fallen or hasn't moved in a while

Used responsibly, AI can support aging in place and empower users with limited mobility. Many of these features are now standard in mainstream devices, not just specialized medical tools.

---

7. The Future of Smart Living

What's next for AI at home?

- **Predictive environments**: Your home adjusts based on your mood, weather, or calendar.
- **Emotion-aware or context-aware systems**: AI may detect stress or fatigue and offer music, lighting, or temperature adjustments.
- **Home robots**: Beyond cleaning, robots may help with cooking, caregiving, or companionship.

- **Interconnected ecosystems**: Multiple devices working together in seamless orchestration, adapting to life changes over time.

The goal isn't to turn your home into a robot butler — it's to create a responsive environment that supports your lifestyle.

Quick Takeaways: AI at Home

» Smarter Living

AI helps your home adapt to your habits, routines, and preferences over time.

» Devices with Brains

From thermostats to vacuums, many devices use machine learning to optimize performance.

» Privacy Matters

Be aware of what your devices are listening to, watching, and sending to the cloud.

» Start with Intention

Smart homes work best when you set thoughtful routines and boundaries.

» Not Just for Comfort

AI can also enhance safety, accessibility, and independence.

Chapter Summary

AI is turning houses into responsive environments that adjust to your comfort, safety, and preferences. From thermostats that learn your schedule to security systems that alert you in real time, smart home devices offer real benefits — if used intentionally. While the convenience is clear, privacy and security remain critical concerns. By starting small, customizing features, and understanding what your devices actually do, you can use AI at home to live more efficiently, more safely, and more comfortably. The key is intentional setup: a smart home should adapt to you, not the other way around.

Chapter 19

AI IN BUSINESS AND WORK PRODUCTIVITY, MANAGEMENT, AND DECISION MAKING

Introduction: Smarter Workplaces

Whether you're replying to emails, managing a team, or analyzing trends, chances are AI has already touched your workflow. From background tools that summarize meetings to high-level decision engines that drive business strategy, AI is rapidly becoming an operational collaborator — supporting how we work without replacing human judgment.

This chapter explores how AI is transforming productivity, communication, hiring, and management across sectors. We'll look at both the benefits and the boundaries, and ask a deeper question: Is AI your coworker or your boss?

---

1. AI as a Personal Productivity Assistant

AI is becoming an indispensable tool for individual productivity, not just for executives and knowledge workers but for anyone with a digital workflow. Think of it as a digital aide that handles routine tasks, manages information overload, and frees up mental bandwidth.

Email and Communication

Email tools like Gmail Smart Compose suggest sentences as you type, learning your tone and frequently used phrases. Superhuman AI and Outlook's Copilot go further, summarizing long threads, prioritizing messages, and even recommending responses. Some systems can draft entire replies for approval or personalize outreach emails at scale.

These tools don't 'understand' email content in a human sense — they identify patterns based on similar messages they were trained on.

Calendar and Scheduling

AI scheduling assistants like x.ai, Reclaim.ai, or Motion optimize meetings based on availability, urgency, and work-life balance. Some can rearrange your calendar in real time based on shifting priorities—ensuring you stay focused on high-impact work.

Knowledge Management and Summarization

Tools like Notion AI, Otter.ai, and Fireflies capture meetings, convert them into searchable notes, and extract action items. Others, like Mem or Reflect, automatically organize your notes and link ideas across projects.

Writing and Editing

From Grammarly and Jasper to Wordtune and ChatGPT, AI writing tools enhance clarity, adjust tone, and generate first drafts for everything from emails to reports and blog posts.

These systems enhance clarity and tone by learning from examples of effective writing, not by applying subjective judgment.

Visual Support

Even slide creation is being AI-assisted. Tools like Beautiful.ai or Gamma.ai help generate polished presentations with minimal effort.

► *The goal isn't just speed—it's focus. By handling low-value tasks, AI helps you reclaim time for strategic thinking, creativity, or rest.*

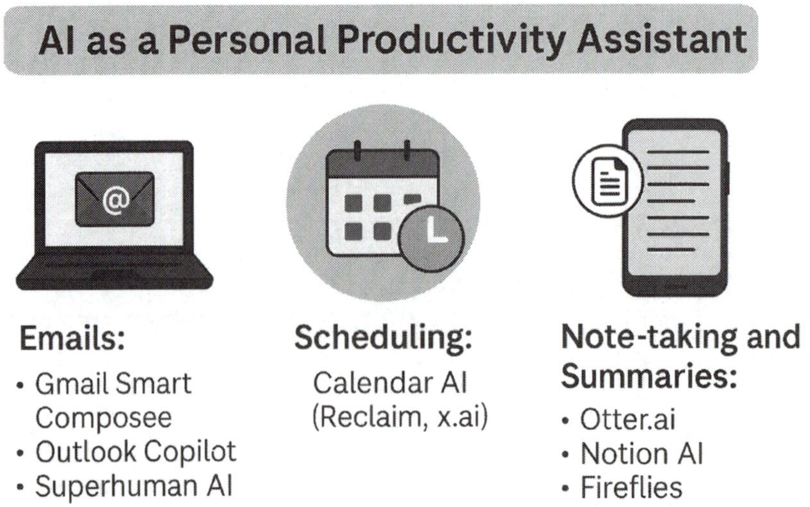

Example: With Microsoft Copilot integrated into Word or Excel, you can summarize a long report, create a presentation outline, or generate a budget projection in seconds.

---

2. Task Automation and Workflow Efficiency

AI doesn't just assist individuals — it automates business operations across entire departments. The focus is no longer just on cost-cutting, but on unlocking hidden value, improving customer satisfaction, and enabling scalability. By streamlining repetitive tasks and integrating real-time data insights, AI allows teams to make faster, more informed decisions. However, AI-driven recommendations depend on the quality and completeness of the underlying data — a point often overlooked when automating workflows. It reduces human error, enhances service personalization, and frees up talent to focus on innovation rather than administration. As workflows become more intelligent and adaptive, AI becomes not just a tool for efficiency — but a strategic driver of business transformation.

CRM Automation

AI-powered CRMs like Salesforce Einstein and HubSpot AI optimize lead scoring, automate follow-ups, and suggest the best time to reach out based on past behavior. These systems reduce manual data entry and help sales teams close more deals with less guesswork.

Data Analytics and BI

Modern dashboards (e.g., Tableau Pulse, Power BI with Copilot, ThoughtSpot) go beyond charts. They surface insights proactively — flagging outliers, forecasting trends, and generating narrative summaries. This shifts the role of analysts from report-builders to decision facilitators.

Document Intelligence

AI tools like UiPath, ABBYY, and Rossum automate the handling of complex documents — reading invoices, extracting key details, filing them correctly, and notifying relevant stakeholders. They dramatically cut time and human error in back-office tasks.

AI in Logistics and Supply Chains

Predictive models help optimize routes, forecast delays, manage inventory, and ensure smoother operations in warehousing, transport, and procurement.

► *Automation isn't just about speed—it's about precision, consistency, and enabling human teams to focus on higher-order challenges. Well-designed systems keep humans involved at key decision points, especially when uncertainty is high.*

Example: A mid-size logistics firm used UiPath to automate invoice processing, reducing turnaround from 3 days to 4 hours — freeing staff to focus on client service.

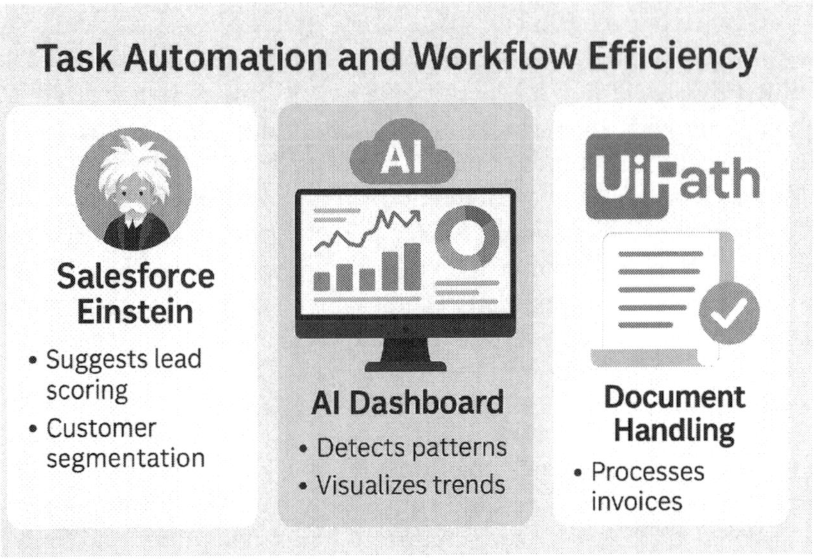

---

3. Human Resources and Hiring

AI is transforming how companies attract, evaluate, onboard, and retain talent — but it comes with serious caveats.

Recruiting at Scale

AI recruitment tools screen thousands of resumes in seconds, prioritizing candidates based on qualifications, experience, and keyword matches. Platforms like HireVue, Pymetrics, or Lever add features like AI interview scheduling and sentiment scoring.

These systems should never be used without fairness checks, because any patterns learned from historical data can unintentionally reinforce inequality.

Candidate Experience

AI-powered chatbots guide applicants through the hiring process, answer FAQs, and provide status updates — improving communication and reducing drop-offs.

Onboarding Automation

Virtual onboarding agents introduce new hires to company tools, policies, and workflows. They can personalize the experience based on the employee's role or location, helping them get up to speed quickly.

Performance Monitoring and Engagement

Some companies use AI tools to monitor productivity, track work hours, and even analyze tone in communication tools like Slack. Others apply AI to detect disengagement or burnout based on behavioral signals.

► **Ethical Risk Alert**: Using historical hiring data can reproduce past biases—disadvantaging women, minorities, or nontraditional candidates. Transparency, auditability, and human oversight are essential.

► *AI in HR should enhance fairness and efficiency, not become a black box that makes irreversible decisions.*

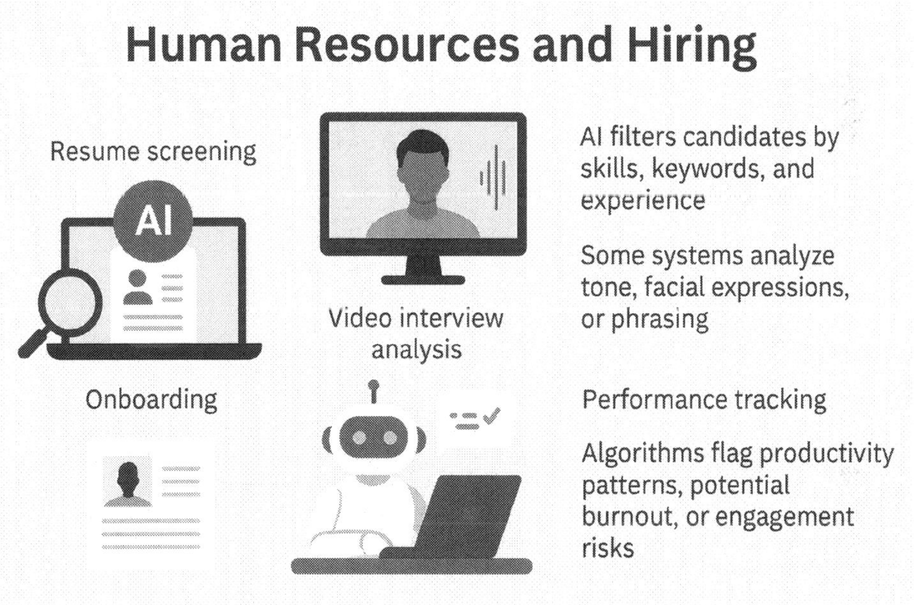

4. Collaboration and Teamwork

Collaboration is no longer limited to shared documents and video calls. AI is reshaping how teams ideate, plan, and communicate — especially in hybrid and global work environments.

AI for Brainstorming and Ideation

Generative AI models like ChatGPT or Gemini can help teams brainstorm product names, campaign slogans, or new features. They generate possible directions by remixing patterns from the data they were trained on — they do not provide original insights in the human sense. Rather than replacing creative teams, they offer starting points and unblock stuck conversations.

Project and Task Management

Platforms like Asana, Trello, Monday.com, and ClickUp now include AI assistants that:

- Predict which tasks are likely to slip
- Auto-suggest assignments based on workload
- Generate progress updates or sprint retrospectives

Cross-Team Communication

AI-based translation, summarization, and sentiment analysis tools help diverse teams collaborate clearly and respectfully across time zones and languages. Grammarly Business and Wordtune Teams help ensure tone and clarity in professional writing.

Virtual Meeting Support

Real-time transcription tools (like Otter.ai or Microsoft Teams' AI features) create searchable meeting notes, tag action items, and send summaries — ensuring no one is left out, even if they miss the meeting.

► *AI improves collaboration by reducing friction—turning unstructured conversations into organized, actionable insights.*

Use Case: In remote teams, AI may help level the playing field by capturing everyone's ideas and translating messages clearly.

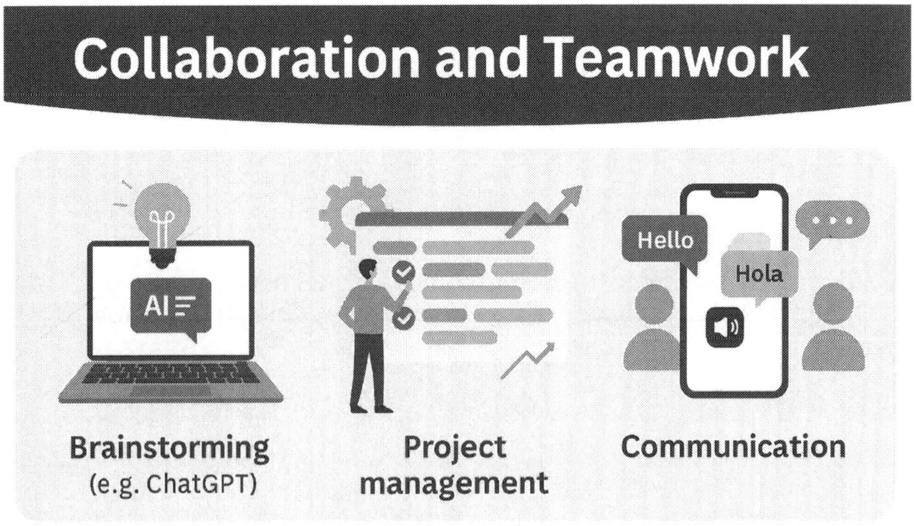

5. AI as Coworker vs. AI as Overseer

There's a subtle shift happening in workplaces: AI isn't just a helper—sometimes, it's a monitor or even a manager.

- **As Coworker**: AI collaborates with you. It drafts, supports, calculates, or suggests. You remain in control.
- **As Overseer**: AI monitors keystrokes, flags anomalies, tracks attendance, or even scores your performance.

This distinction matters:

- **Trust and Morale**: Workers may embrace AI that supports them but feel anxiety or resistance toward surveillance.
- **Transparency**: It's crucial to explain what AI systems do and how decisions are made.
- **Fairness and Accountability**: Who takes responsibility when an AI-driven system makes a wrong call?

When to Trust AI — and When to Step In

Not every decision is best handled by AI alone—or by humans. A growing concept in workplace design is **augmented intelligence**: combining the strengths of both.

For example, in fraud detection, AI systems are excellent at flagging clear-cut cases when they have high confidence — whether it's a false alarm or a likely fraud. But when the case is ambiguous and the AI is uncertain, **humans tend to outperform**. Ambiguity exposes the limits of pattern recognition systems, especially when context or nuance matters. They can pull in additional context, double-check assumptions, or ask questions the AI cannot.

This creates a zone of ideal collaboration:

AI handles what it's sure of; humans review what's unclear.

Design Matters

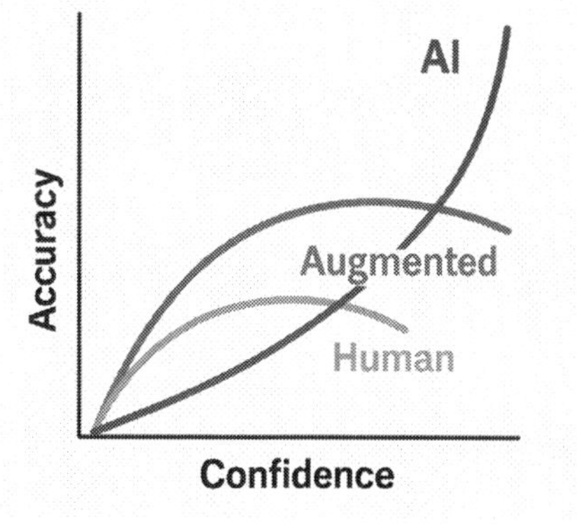

The way AI suggestions are presented affects how people respond:

- **Forced display** (showing the AI's opinion upfront) can cause **automation bias**, where users rely too heavily on the machine — even when it's wrong.

- **Optional display** (letting the human decide when to see the AI's suggestion) leads to better judgment because the user forms their own opinion first.

Transparency Can Backfire

Surprisingly, when an AI shares its confidence score (e.g., "79% likely to be correct"), people tend to **trust it less**—even when it's statistically accurate. We're more likely to follow an overconfident AI than a humble one, even if the latter is more honest. This effect — automation complacency — increases risk even when the model is accurate overall.

In workplaces where AI acts as a collaborator, the **timing, format, and framing of AI input** can shape decision outcomes just as much as the algorithm itself. Design choices are human decisions, too.

Thought Prompt: If your promotion or workload is decided by an algorithm, should you have the right to appeal or understand how it works?

Second Thought Prompt: What makes you *trust* an AI's suggestion? Its past accuracy? Its confidence score? Or would you rather form your own opinion first?

---

6. Risks and Challenges

While AI increases efficiency, it also introduces complex risks:

- **Job displacement**: Some roles may be augmented, others replaced. The impact varies by industry and skill level.
- **Algorithmic bias**: Hiring, lending, and evaluation tools may reinforce unfair outcomes.
- **Data privacy**: AI systems require vast data access—raising questions about surveillance, consent, and misuse.
- **Over-reliance**: When humans stop double-checking AI suggestions, mistakes scale quickly.

Cognitive Bias Is a Risk, Too

Even when AI systems work well, **humans don't always use them well**. We may over-trust machines, or ignore them completely. Both tendencies lead to worse decisions.

- Show an AI recommendation too soon, and people might stop thinking critically.
- Present a confidence score, and users might tune it out — preferring certainty over nuance.

In AI-assisted workplaces, good decision-making depends on more than just good models. It depends on **how we design, explain, and use** those models in practice.

Mitigating these challenges requires not just better tech, but better policies and human oversight.

Clear governance frameworks help define who is accountable for decisions influenced by AI — a crucial requirement as automation expands.

---

7. The Future of Work with AI

We're still in the early days. Expect to see:

- **AI copilots for every role:** Designers, analysts, marketers, and support staff will all have LLM-based assistants.
- **Adaptive workflows:** Systems that optimize your task list or meeting schedule in real time.
- **Augmented decision-making:** AI won't replace managers, but it will reshape how they prioritize, allocate, and strategize.
- **Human-AI teaming:** Future work will rely on knowing what to automate and what to do yourself.

To help guide this transition, businesses are adopting structured frameworks like the IBM AI Ladder, which outlines four key stages:

- **Collect:** Gather high-quality data from multiple sources such as internal databases, documents, or IoT devices.
- **Organize:** Clean, categorize, and govern that data so it becomes reliable and ready for analysis.
- **Analyze:** Apply analytics and machine learning to identify patterns, generate insights, and make predictions.
- **Infuse:** Embed those AI insights into real workflows — such as CRM platforms, inventory systems, or customer-support tools — so the benefits show up directly in day-to-day operations.

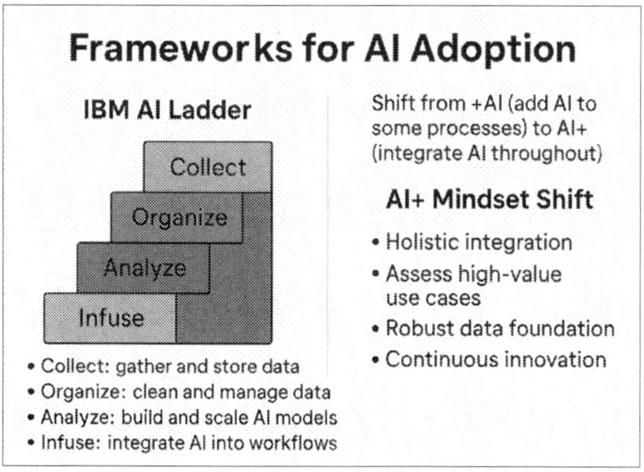

This approach supports not just technical adoption, but cultural change — shifting from "+AI" (adding AI as a tool) to "AI+" (designing workflows around AI from the start).

Business Insight: AI adoption works best when aligned to strategic goals, backed by clean data, and supported by an informed workforce. Continuous monitoring and adaptation ensure long-term success.

The most successful workers and businesses won't just use AI. They'll understand how to collaborate with it — and design their operations to maximize its strengths while preserving human judgment.

Quick Takeaways: AI in Business and Work

» Smarter Personal Productivity

AI helps you write, schedule, summarize, and organize with less friction.

» Workflow Optimization

From customer service to data dashboards, AI makes business processes faster and more informed.

» Hiring and Management Tools

AI is now screening resumes, onboarding employees, and flagging performance trends — but bias is a real risk.

» Coworker vs. Overseer

Helpful AI boosts productivity. Monitoring AI raises ethical and emotional concerns.

» Know the Risks

Effective AI use requires human judgment, evaluation of outputs, and awareness of system limits.

Chapter Summary

AI is becoming a powerful force in the workplace — automating tasks, assisting decision-making, and reshaping how teams collaborate. From smarter email to predictive dashboards to AI-enhanced hiring, businesses are using these tools to move faster and stay competitive. But alongside the gains come questions about fairness, transparency, and the future of work. As AI increasingly becomes both coworker and overseer, the challenge isn't just how to use it, but how to use it well — in ways that support human values, not just business outcomes. In the end, it's not about working more — it's about working smarter, with AI by your side. As with all AI systems discussed in earlier chapters, success depends on clear boundaries, responsible use, and keeping humans meaningfully involved.

Chapter 20

AI IN CREATIVITY
TOOLS FOR WRITERS, ARTISTS, MUSICIANS, AND MORE

Introduction: Machines That Make

For centuries, creativity was seen as uniquely human — a spark of imagination that separated us from machines. But that line is blurring. Today, AI can paint portraits, compose music, write screenplays, and even animate videos. These tools don't "feel" or "imagine" like humans do — but they can remix patterns, generate variations, and help creators explore new possibilities faster than ever before.

The rise of AI in creative work raises both excitement and concern. Are these tools partners in the creative process — or are they replacing the process itself? What does authorship mean when a machine contributes? Can a synthetic painting or melody truly be called art?

This chapter explores the expanding role of AI in creative industries—from writing and design to music and video. We'll look at how the tools work, who's using them, and how they're changing what it means to create.

---

1. Visual Arts: From Prompt to Painting

AI tools like **Midjourney**, **DALL·E**, and **Stable Diffusion** allow anyone to create images with nothing more than a text prompt. You might type "a cat playing violin in the style of Van Gogh" and receive a detailed, stylized rendering in seconds.

Professional tools like **Runway ML** and **Adobe Firefly** take this further with video generation, background replacement, and AI-powered editing.

Use cases include:

- **Graphic designers** generating mood boards or ad concepts
- **Indie filmmakers** creating special effects or storyboards
- **Hobbyists** making art for fun or social sharing

These tools don't replace artistic intent — but they speed up experimentation and reduce production barriers.

Many top companies are now embedding generative AI in creative tools:

- **Adobe Sensei** enhances photo editing and font recognition
- **Google Gemini** powers AI enhancements in Google Photos and music generation via Magenta

---

2. Music: Composing with Algorithms

Music generation tools like **Suno**, **Udio**, **Amper Music**, and **Aiva** let users create original tracks using genre, mood, or instrumentation prompts.

You can use these tools to:

- Generate background music for a YouTube video
- Compose loops or melodies for electronic music
- Create royalty-free soundtracks for games or ads

Professionals may use AI to speed up drafts or fill creative gaps. Beginners might use them to learn structure and play with ideas.

Importantly, the music generated is usually based on pattern learning — not emotional experience. That raises questions about expressiveness and authenticity.

---

3. Writing: From First Draft to Final Edit

AI writing assistants have exploded in popularity:

- **ChatGPT**, **Jasper**, and **Notion AI** help generate blogs, emails, headlines, or even poems.
- **Grammarly** and **QuillBot** enhance grammar, tone, and clarity.
- **Sudowrite** helps fiction writers develop plots or rewrite passages.

Some models, like Claude or Gemini Advanced, even support multi-modal prompts — letting writers include images or video cues to shape output.

Amateur creators use these tools to get started. Professionals use them to draft faster, explore variations, or polish their voice.

But writing is more than just words on a page — it's tone, purpose, and audience. That's where human guidance remains essential.

Even the most advanced models help with language and structure, not with intention, judgment, or lived experience — the uniquely human elements that shape meaningful writing.

---

4. Co-Creation vs. Automation: Who's Really Creating?

There's a spectrum of creativity happening:

- **Assisted Creativity**: You give the ideas; the AI helps shape or polish them.
- **Prompt-Based Generation**: You set the direction; the AI produces the output.
- **Fully Automated Content**: You hit a button and get finished work with minimal input.

This raises a key debate: What counts as "real" art?

Some argue that as long as a human sets the intention, the art is theirs. Others say true creativity involves process, revision, and emotion—things machines don't experience.

The truth may lie in the **collaboration**. Many creators now describe AI as a *muse*, not a replacement. It can inspire new directions, generate variants, or help overcome blocks — but the vision is still human.

---

5. Amateurs vs. Professionals: Different Goals, Different Tools

AI's impact differs based on experience:

- **Amateurs** often use AI to explore. It lowers the barrier to entry, allowing people to write stories, compose songs, or create art without technical skills.
- **Professionals** use AI to accelerate parts of their workflow. They integrate it into concepting, drafting, or editing phases — but still bring expertise, originality, and critical judgment.

In short, amateurs often use AI to create; professionals use AI to enhance, refine, and accelerate their existing skills.

The line is blurring — but the intent and outcome often differ.

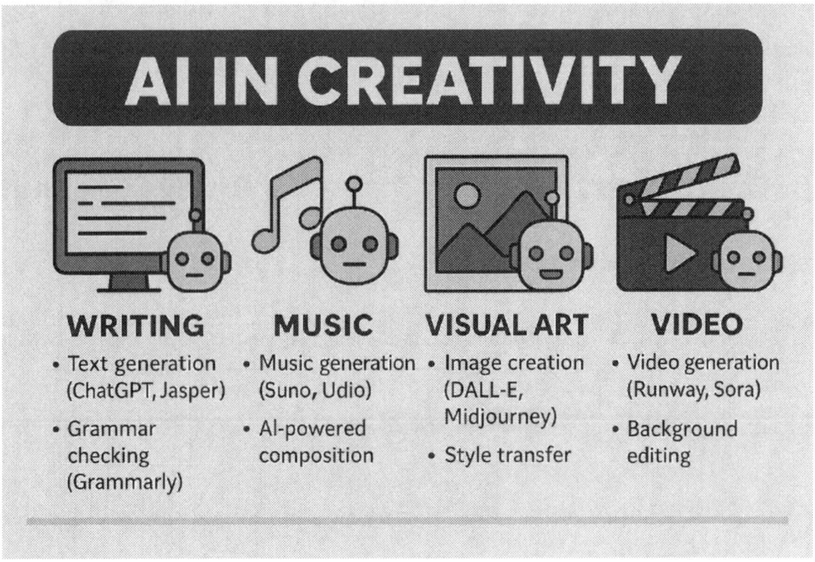

---

6. Legal and Ethical Challenges

AI creativity also raises serious questions:

- **Authorship**: Who owns AI-generated content — the user, the tool's developer, or no one?
- **Plagiarism**: Some AI models generate content that closely resembles their training data. Can that be considered theft?
- **Fair Use and Style Mimicry**: Tools that generate art in the style of a specific artist prompt concern over consent and compensation.
- **Deepfakes and Misinformation**: Synthetic media can be misused to create false realities, posing challenges for trust and verification.

Courts and copyright offices are still catching up. In recent years, the U.S. Copyright Office reaffirmed that content must have "human authorship" to be copyrighted—but that line is increasingly difficult to define.

---

7. Looking Ahead: Creative Futures

As AI evolves, it won't just enhance creative tools — it will redefine what it means to create. We're entering an era where humans and machines co-invent, co-compose, and co-narrate experiences in ways previously unimaginable.

Expect to see developments like:

Personalized Creative Companions

Future AI tools will not just generate generic content — they'll learn your voice, your artistic style, your musical preferences, and your creative habits. Imagine an AI painter that evolves alongside you, or a music partner that riffs with your emotions. These personalized systems could suggest improvements, generate complementary work, or even evolve their "personality" to mirror your artistic journey.

Interactive and Living Stories

AI-generated storytelling will become non-linear, personalized, and interactive. Readers could influence the plot in real time, making each experience unique. AI dungeon masters, choose-your-own-adventure companions, or co-writers could allow readers to become participants — blurring the line between audience and author.

Immersive and Generative AR/VR

In augmented and virtual reality, AI will power responsive environments that react to your mood, gestures, and narrative choices. A future artist might not draw on a screen — but sculpt light and color in a virtual 3D space, with AI helping interpret intention and filling in details. Musicians may perform alongside AI generated visuals that respond to rhythm, pitch, and tempo.

Collective AI-Created Works

Crowdsourced creativity will take new forms. Platforms may emerge where thousands contribute ideas, prompts, fragments, or annotations — and AI weaves it all into evolving pieces of art, literature, or sound. These "living artworks" may never be finished, constantly growing through community input and AI refinement.

Cross-Modal Synergy

AI will increasingly unify multiple creative domains. For example, a writer might feed a story draft into an AI tool that automatically composes background music, designs illustrations, and generates a voiceover in the tone of the narrative. These tools will remove traditional boundaries between mediums — encouraging hybrid, multimodal expression.

Emotionally Aware Creativity

Next-generation models may sense emotional intent through voice, facial expression, or written tone — adjusting their creative output accordingly. This could lead to more intuitive tools that "feel" your frustration or enthusiasm and adjust the pacing, tone, or style of collaboration in response.

Ethical Frameworks for AI Creators

As generative tools become more capable, frameworks will likely emerge to help creators disclose AI involvement, track content provenance, and ensure credit where it's due. Digital provenance tools and watermarking systems may support this, enabling transparency and fair compensation in hybrid human-AI creative work.

Ultimately, the future of AI in creativity isn't about replacing artists—it's about expanding the creative canvas. AI will take on the routine, the repetitive, and the exploratory—freeing humans to focus on the emotional, intentional, and meaningful aspects of creation.

Quick Takeaways: AI in Creativity

» **Instant Visuals**

AI tools like Midjourney and DALL·E turn text into images in seconds — great for concepts, mood boards, and illustration.

» **Sonic Prototyping**

Music generators like Suno and Udio make composing accessible, even for non-musicians.

» **Writing Partners**

LLMs help with drafting, editing, ideating, and polishing. They speed up — but don't replace — the writing process.

» **Co-Creation, Not Replacement**

AI is best used as a creative partner, not a substitute for originality or vision.

» **Legal and Ethical Gray Zones**

Questions around authorship, ownership, and plagiarism are unresolved — and critical to understand.

» **Beginners and Pros Use AI Differently**

Amateurs explore with it. Professionals integrate it. The tool is the same — the goals are not.

Chapter Summary

Artificial intelligence is transforming the creative process across disciplines — writing, music, visual arts, and more. Tools like Midjourney, ChatGPT, Udio, and Grammarly let creators generate, iterate, and polish faster than ever before.

But AI doesn't create meaning. It generates output based on patterns. Meaning comes from human intention — the emotional, cultural, and experiential layers that AI cannot replicate. That means the most powerful results come when humans lead — and AI supports. For amateurs, these tools open new doors. For professionals, they offer a boost in speed and variation.

The big questions — about authenticity, authorship, and ethics — are still unfolding. But one thing is clear: AI is no longer just a tool for analysis. It's becoming a collaborator in imagination.

Chapter 21

AI IN HEALTHCARE
DIAGNOSING, MONITORING, AND SUPPORTING WELLNESS

Introduction: Intelligence That Cares

AI isn't just transforming emails or music playlists — it's helping save lives. In hospitals, clinics, labs, and homes, AI is improving how we detect disease, tailor treatments, and support well-being. From scanning X-rays to predicting heart conditions, AI brings speed, consistency, and insight to one of the most vital sectors of human life: healthcare.

But trust matters. When your diagnosis, treatment, or safety depends on a machine's decision, it raises the stakes. That's why AI in healthcare must remain a tool for clinicians, never an independent authority. This chapter explores how AI is being used to support patients, doctors, and researchers—while also highlighting the limitations, risks, and open questions we must confront.

---

1. Diagnosing with Precision

AI has become a powerful tool in medical imaging:

- **X-rays and MRIs**: Tools like Aidoc and Zebra Medical analyze scans faster than human radiologists in some cases — highlighting fractures, tumors, or blockages.
- **Retinal Scans**: Google's health research teams have built AI systems that help detect diabetic retinopathy and other eye diseases with high accuracy.
- **Skin Lesion Analysis**: Apps like SkinVision and AI dermatology platforms analyze moles and rashes to flag signs of melanoma or other conditions.

These systems don't replace doctors — but they can act as a second set of eyes, improving detection rates and reducing human error.

2. Personalized Medicine and Risk Prediction

AI models can combine genetic, lifestyle, and historical data to:

- Predict a patient's risk for diseases like diabetes, cancer, or heart conditions
- Suggest personalized treatment plans based on likely outcomes
- Adjust medication dosages for optimal effectiveness and safety

This is part of a growing field known as **precision medicine**, where care is tailored to the individual, not just the population.

---

3. AI in Drug Discovery

Developing a new drug typically takes over a decade and costs billions of dollars. AI is helping speed up that process:

- **Molecular design**: Platforms like Insilico Medicine and tools like AlphaFold help predict protein structures and guide researchers toward promising drug targets by modeling how those structures interact.
- **Target discovery**: AI scans biological databases to find new pathways and potential disease mechanisms.
- **Preclinical analysis**: Algorithms simulate how new compounds might behave in the body — cutting down on early trial failures.

This doesn't eliminate human research — it amplifies it by reducing guesswork and focusing attention.

---

4. Optimizing Clinical Trials

AI helps improve clinical trial efficiency by:

- Identifying suitable patient candidates based on medical records and genetic profiles
- Predicting trial outcomes or dropouts based on previous data
- Monitoring adherence and side effects in real time through wearables or mobile check-ins

Smarter trials mean faster approvals, lower costs, and potentially life-saving treatments reaching patients sooner.

5. Mental Health and Virtual Assistants

AI chatbots like **Woebot** and **Wysa** use cognitive behavioral therapy (CBT) techniques to support people dealing with stress, anxiety, or loneliness.

These tools:

- Offer a nonjudgmental space for reflection
- Deliver guided exercises or coping strategies
- Are available 24/7 — even when human therapists aren't

While not a replacement for licensed care, they help bridge gaps in accessibility and affordability. They are best used for early support or mild symptoms, not for crises or complex clinical needs.

---

6. Wearables and Real-Time Monitoring

Smartwatches and fitness bands are evolving toward medical-grade capabilities:

- Track heart rate variability, oxygen saturation (SpO_2), fertility, and sleep cycles
- Detect irregular heart rhythms (like atrial fibrillation) and alert users or physicians
- Help manage chronic conditions like diabetes or asthma

Platforms like Apple Health, Fitbit, and Oura increasingly integrate AI to personalize insights and spot long-term trends.

---

7. Challenges: Bias, Liability, and Trust

While AI shows promise, its use in healthcare must be cautious and deliberate.

Key issues include:

- **Bias in training data**: If datasets underrepresent certain populations (e.g., minorities or women), AI models may perform poorly or unfairly for those groups.
- **Accountability**: If an AI misdiagnoses a patient, who is responsible — the doctor, the hospital, or the software maker?
- **Privacy and consent**: Healthcare data is sensitive. Users must trust that their information is secure and used ethically.
- **Explainability**: Doctors and patients often want to know *why* a diagnosis was made. Black-box algorithms can undermine confidence in AI decisions.

These aren't just technical problems—they're ethical and social questions that must be addressed through regulation, oversight, and public dialogue.

---

8. The Future of AI in Healthcare

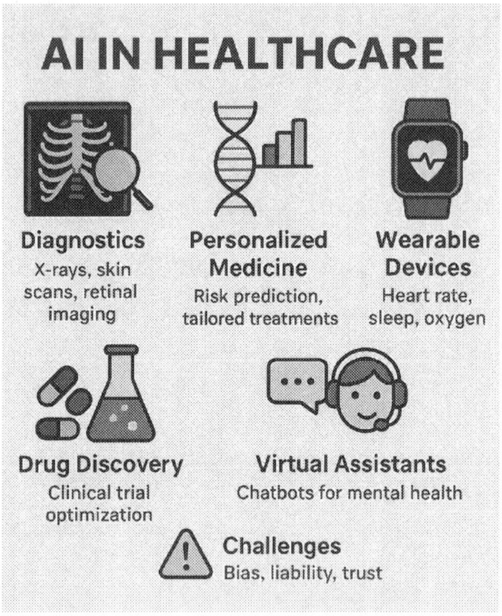

As AI continues to mature, its role in healthcare is moving from assistance to integration — becoming an invisible, intelligent layer embedded within daily care routines, medical research, and patient relationships. The most exciting developments on the horizon focus not just on efficiency, but on **enhancing care quality, equity, and personalization at scale**.

Ambient AI in the Exam Room

Future medical consultations may be accompanied by AI that listens passively — not to judge, but to support. These systems can:

- Transcribe and summarize doctor–patient conversations in real time
- Auto-suggest diagnoses, treatment plans, or follow-up actions
- Generate patient instructions or update electronic health records (EHRs)

This "clinical co-pilot" relieves physicians from tedious documentation, allowing more face-to-face time with patients — and reducing burnout.

Digital Twins for Personalized Care

A digital twin is a dynamic, data-driven model of an individual's health profile. It simulates how **you** — not just someone statistically similar — might react to medications, surgeries, or lifestyle changes.

These models are built using a combination of:

- Genomic data
- Medical history
- Real-time biometric inputs (e.g., wearables)

Doctors could use your digital twin to run virtual experiments before making real-life decisions, reducing trial-and-error and improving safety.

AI for Health Equity and Global Access

AI has the potential to democratize healthcare — if designed inclusively:

- **Language translation and local dialect processing** can bring health access to underserved linguistic communities.
- **Mobile diagnostic tools** powered by AI can serve rural areas without access to specialists.
- **Low-cost image analysis apps** could help detect conditions like tuberculosis, anemia, or skin infections using a smartphone camera.
- Major nonprofits and institutions are now investing in AI to **bridge care gaps** rather than widen them — an essential step in ensuring that innovation benefits all.

Predictive, Preventative, and Proactive Care

We're shifting from treating illness to predicting and preventing it. Future AI systems will:

- Spot **subtle physiological changes** before symptoms arise
- Recommend **lifestyle interventions** tailored to individual risk factors
- Alert healthcare providers when intervention is urgently needed

Imagine an AI that flags a high-risk pregnancy complication two weeks before it would otherwise be noticed, or detects early signs of relapse in mental health patients through subtle shifts in text or voice patterns.

Multi-modal, Continuous Monitoring

AI will increasingly combine data streams — imaging, lab tests, EHRs, wearables, and even patient voice or facial expressions — to build a **comprehensive, real-time picture** of someone's health. While these systems are still emerging, early research shows strong potential for earlier detection and more proactive care.

Such holistic monitoring could lead to:

- Earlier detection of complex diseases
- Better management of chronic conditions
- Reduced hospital readmissions

Regulatory and Ethical Maturity

For these futures to succeed, AI governance must evolve in parallel. Expect:

- **Global standards** on explainability, data privacy, and bias mitigation
- **Certification of medical AI tools** (like FDA or CE approval)

- **Transparency dashboards** showing how AI systems perform across diverse populations

As these guardrails strengthen, public trust — a critical factor in healthcare — can grow accordingly.

Quick Takeaways: AI in Healthcare

» Faster, Smarter Diagnoses

AI helps doctors detect issues in scans, images, and data with increasing accuracy.

» Tailored Treatments

Predictive tools personalize care based on genetics, risk factors, and outcomes.

» AI in the Lab

Drug discovery and clinical trials are being accelerated with pattern recognition and data-driven predictions.

» Mental Health Support

AI chatbots offer accessible emotional care, especially when human therapists aren't available.

» Real-Time Health Monitoring

Wearables track vital signs and help manage chronic conditions more proactively.

» Caution Required

Bias, explainability, and liability remain open challenges. AI should support—not replace — human care.

Chapter Summary

From diagnosing disease to managing wellness, AI is changing how healthcare works. It's speeding up diagnosis, optimizing treatment, and even discovering new drugs. Tools like AI-powered imaging, wearable trackers, and therapeutic chatbots offer new levels of insight and support.

But these gains come with complexity. Medical decisions are high-stakes and deeply personal. We must ensure that AI systems are fair, trustworthy, and accountable. The future of healthcare won't be human *or* machine — it will be human *with* machine, working together to make care smarter, faster, and more accessible.

As with all powerful tools, the key is not just capability but governance — ensuring AI supports clinical expertise rather than substituting for it.

Chapter 22

SELF-DRIVING CARS – THE AI BEHIND THE WHEEL

Introduction: From Hands on the Wheel to Hands-Free Travel

Imagine sitting in your car, sipping coffee, reading the news, and letting the vehicle handle everything — from lane changes to parking. That's the promise of self-driving technology. But while the dream of fully autonomous cars captures headlines, the road to autonomy is complex, full of technical challenges, ethical questions, and gradual steps.

AI is already transforming how we drive, even before full autonomy. From adaptive cruise control to emergency braking and lane assist, many cars today are semi-autonomous, helping drivers stay safer and more efficient. This chapter explores the tech behind these systems, how full autonomy is being developed, and what it means for our cities, jobs, and daily lives.

---

1. How Self-Driving Cars Work

Autonomous vehicles rely on a combination of sensors, algorithms, and real-time decision-making systems. Key components include:

- **Sensors:** Cameras, LiDAR (laser scanning), radar, and ultrasonic sensors collect data about the vehicle's surroundings.
- **Sensor Fusion:** AI combines these data sources into a unified 3D model of the environment. For example, radar may detect speed and distance, while cameras interpret traffic signs and pedestrians.
- **Perception and Prediction:** The system identifies objects (cars, cyclists, signs), predicts how they'll move, and plans its response.
- **Control Systems:** AI translates decisions into physical movement—steering, braking, accelerating—with rapid precision.

This fusion creates an automated driving system that can perceive its surroundings, plan actions, and control the vehicle — though not with human-like understanding or common sense.

2. Assisted Driving Before Autonomy

While fully self-driving cars remain in development, assisted driving is already here. These features support, not replace, driver responsibility — the human must remain attentive and in control:

- **Driver Assistance Systems (ADAS):** Include lane-keeping assist, blind spot detection, adaptive cruise control, and collision warnings.
- **Autopilot Features:** Tesla's Autopilot and GM's Super Cruise offer hands-free highway driving under supervision.
- **Real-Time Detection:** AI analyzes objects and road conditions faster than humans, reducing the chance of human error due to distraction or fatigue.

These technologies offer tangible safety and convenience benefits — especially in situations where human attention is limited. Think of AI as your extra set of eyes (and reflexes) on the road.

---

3. Levels of Driving Autonomy

The **SAE (Society of Automotive Engineers)** defines six levels of autonomy:

Most commercial systems today are at Level 2. Level 2 means the vehicle can assist with steering and speed, but the driver must supervise continuously. Higher levels are being tested, but are not yet ready for mass deployment.

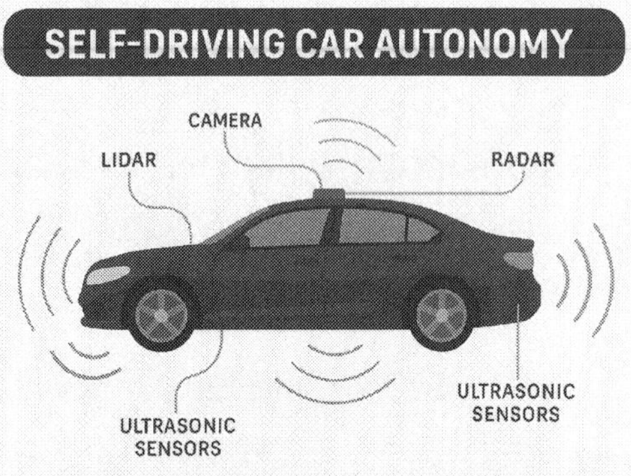

LEVEL	DESCRIPTION	HUMAN INVOLVEMENT
L0	No automation	Human
L1	Driver assistance	Human with assist
L2	Partial automation	Human monitors
L3	Conditional automation	Human if required
L4	High automation	No human in some areas
L5	Full automation	No human required

---

4. Leading Players

Several major companies are racing toward full autonomy:

- **Tesla:** Uses a camera-based system and neural networks marketed as "Full Self-Driving," but still classified as Level 2 because the human must remain responsible.

- **Waymo (Alphabet/Google):** Operates fully driverless taxis in parts of Phoenix, with a strong emphasis on LiDAR and safety redundancy.
- **Cruise (GM):** Also operates autonomous taxis and focuses on urban driving.
- **Aurora, Zoox, and Mobileye:** Building systems for freight, delivery, and ridesharing.

Each company balances safety, scalability, and cost in different ways.

---

5. Safety, Edge Cases, and Ethics

Self-driving systems must handle not only everyday driving but rare and unpredictable events — known as **edge cases**:

- A pedestrian on a scooter weaving through traffic
- An unexpected road closure
- A child running into the street chasing a ball

AI must make split-second decisions, sometimes with no perfect outcome. This leads to famous **ethical dilemmas** like the "trolley problem": Should the car prioritize passengers or pedestrians in a no-win crash scenario?

In practice, engineers focus less on philosophical scenarios and more on avoiding such no-win situations through defensive planning.

Building trust in these systems depends on transparent reasoning, fail-safes, and rigorous testing.

---

6. Regulation and Adoption Timelines

Self-driving regulation varies by country and region. Key points include:

- **Testing Permits:** Cities like San Francisco and Phoenix allow AV testing on public roads.
- **Safety Standards:** Still emerging. Governments must decide how to certify a vehicle that "thinks."
- **Liability:** Who is responsible in a crash — driver, manufacturer, or software developer?
- **Timeline:** Experts differ widely, but most agree widespread Level 4 or 5 autonomy is still many years away.

Progress is steady but cautious — regulators want proof that autonomy reduces, not increases, risk.

7. Broader Impact

Self-driving cars won't just change how we drive — they'll reshape our world:

- **Urban Planning:** With fewer parking needs, cities could redesign roads and free up real estate.
- **Labor Markets:** Long-haul trucking and taxi services could be disrupted. Retraining may be needed for millions.
- **Environment:** Autonomous fleets may be electric and more efficient, lowering emissions. But more cars could also mean more traffic.

The social, economic, and environmental effects will be felt far beyond the car industry.

Quick Takeaways: Self-Driving Cars

» AI on the Road

Self-driving cars use sensors, computer vision, and deep learning to navigate safely and intelligently.

» Help First, Autonomy Later

Assisted driving features like lane assist and adaptive cruise control are already making roads safer.

» Not All Self-Driving Is the Same

From Level 0 to Level 5, autonomy comes in degrees — most systems today still require human oversight.

» The Big Names

Tesla, Waymo, Cruise, and others are leading the charge with very different tech stacks.

» Ethics and Edge Cases

Autonomous systems must make decisions in uncertain, morally complex scenarios — requiring transparency and accountability.

» Societal Shifts Ahead

Self-driving tech will impact urban design, jobs, safety standards, and how we think about mobility.

Chapter Summary

Self-driving vehicles are one of the most ambitious uses of AI — combining perception, planning, and action in real-time. While full autonomy (Level 5) is still a work in progress, assisted driving features already save lives and reduce driver burden. Companies like Tesla, Waymo, and Cruise are advancing the field, but adoption hinges on regulation, safety, and public trust.

AI gives cars more than just speed — it gives them attention, precision, and logic. But in a world of unpredictable people and edge cases, collaboration between human drivers and machine intelligence may remain the safest route forward — for now.

In the near term, the safest future is shared control — AI handling precision tasks, humans providing judgment and oversight.

Chapter 23

BEYOND THE EVERYDAY
WHERE AI IS MAKING QUIET WAVES

Introduction: AI's Expanding Reach

Throughout this part of the book, we've explored how AI shows up in your phone, your job, your car, and your home. But AI doesn't stop at consumer tools. It's also being deployed in classrooms, courtrooms, forests, and climate labs. These uses may not be as visible or personalized, but they could shape our shared future in profound ways.

This chapter takes a brief look at three growing areas where AI is starting to make an impact: education, the environment, and governance. These aren't just add-ons — they represent AI's deeper integration into the systems that structure society.

---

1. AI in Education — Teaching that Adapts to You

AI is transforming how we teach, learn, and assess progress:

Adaptive Learning Platforms

Tools like Squirrel AI or Khanmigo personalize lessons based on how students perform. Struggling with fractions? The system slows down. Moving quickly through algebra? It adjusts accordingly.

AI Tutors and Feedback Assistants

Apps like Duolingo's AI companion or homework-support tools help learners practice skills, clarify concepts, and get real-time feedback.

Grading and Assessment

Platforms such as Gradescope help score quizzes and even open-ended responses using AI-assisted review. These tools reduce administrative load so teachers can focus on instruction.

Concerns include:

- **Bias in AI scoring** that may disadvantage students whose writing style or learning pattern differs from the data the system was trained on.
- **Over-reliance on AI tutors**, which may reduce opportunities for peer discussion and human interaction.

Still, in large classrooms, remote schooling, or areas with limited teacher access, AI can meaningfully expand support.

---

2. AI and the Environment
Modeling a More Sustainable World

AI is becoming a key tool in the fight against climate change and biodiversity loss:

Climate Forecasting and Modeling

AI systems analyze large datasets to improve predictions about temperature shifts, storm intensity, and long-term environmental changes.

Resource Optimization

Smart grids and energy systems use AI to balance load, reduce emissions, and integrate renewable sources more efficiently.

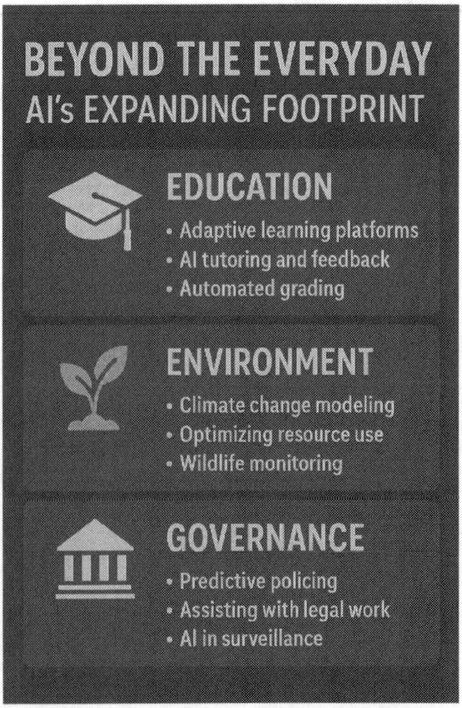

Wildlife and Ecosystem Monitoring

Conservationists use AI to analyze camera trap footage, identify species, track migration, and detect illegal deforestation from satellite imagery.

Examples include:

- Microsoft's **AI for Earth**, supporting projects in agriculture, water management, wetlands, and climate.
- **Rainforest Connection**, which uses AI to detect illegal logging by analyzing audio for chainsaw patterns.

Risks and limitations:

- **Data Gaps:** Under-monitored regions may produce incomplete or biased environmental models.
- **Tech-first Bias:** AI solutions may overlook local expertise or low-tech interventions that are equally effective.

Even with these challenges, AI offers powerful tools to predict, protect, and respond more effectively to environmental pressures.

---

3. AI in Governance and Law Justice and Control in the Algorithmic Age

Governments and legal institutions are beginning to adopt AI tools — sometimes cautiously, sometimes rapidly:

Predictive Policing

Some cities have piloted tools that predict potential crime locations based on historical data. These systems are controversial because crime data often reflects existing societal biases.

AI Legal Assistants

Tools like Casetext or Harvey help lawyers and courts summarize case law, review contracts, or draft memos — improving speed and consistency.

AI in Surveillance

Facial recognition and behavioral analytics are increasingly used in public spaces and transportation hubs, raising concerns about privacy, civil liberties, and overreach.

Challenges are significant:

- **Algorithmic Bias:** Predictive policing and automated risk assessments may reinforce systemic inequalities.
- **Transparency:** Some AI-assisted legal or administrative decisions are opaque, making appeals difficult.
- **Accountability:** If an algorithm affects sentencing, bail, or loan approvals, determining responsibility becomes complex.

Because these systems influence rights and freedoms, they require the strictest oversight, auditability, and public trust.

4. Conclusion: The Expanding Footprint of AI

As AI spreads beyond productivity tools and entertainment, it's beginning to influence how we learn, govern, and protect the planet. These changes bring enormous potential — and equally important questions.

Whether improving educational outcomes, conserving fragile ecosystems, or supporting legal workflows, the power of AI lies not just in what it can do, but in how we choose to use it. These aren't merely technical challenges. They're deeply human ones.

Quick Takeaways: Beyond the Everyday

» **Education**

AI personalizes learning and eases teacher workloads, but must be used responsibly to avoid bias or over-reliance.

» **Environment**

AI powers climate modeling, energy efficiency, and wildlife protection—offering new ways to safeguard the planet.

» **Governance**

AI supports legal research and administration, but also raises serious concerns around surveillance, fairness, and accountability.

Chapter Summary

AI is expanding into domains where decisions affect not just individuals, but whole communities. In classrooms, it adapts lessons to students. In ecosystems, it monitors change and protects biodiversity. In government and law, it offers speed and insight — but demands the highest levels of transparency and fairness.

This chapter closes Part 3 by showing that AI isn't only about convenience or productivity. It's about systems, values, and responsibility. And its real power lies in how wisely — and how ethically — we choose to use it.

PART IV THE ETHICS & FUTURE OF AI

Introduction: Revisiting Critical Questions

By now, we've seen AI in action — from how it works under the hood to how it shows up in your phone, your workplace, your home, and the systems around you. Along the way, we've touched on essential ethical concerns: fairness in hiring tools, the risks of deepfakes, how autonomous systems make decisions, and who is accountable when things go wrong.

In this final section, we return to those issues and look at them more closely. Because while AI's capabilities are impressive, its impact depends entirely on how we design, deploy, and govern it. Ethics, transparency, and long-term thinking aren't side topics — they're the foundation for building AI that benefits everyone.

Part IV explores the challenges, trade-offs, and responsibilities that come with powerful systems. It asks not just what AI can do, but what it should do. And it prepares us to think critically about the choices that will shape the future of AI — and, by extension, the future we all share.

Chapter 24

BIAS AND FAIRNESS IN AI — CAN MACHINES BE FAIR?

Introduction: When Algorithms Discriminate

AI systems are often seen as neutral, mathematical engines — incapable of prejudice. But in reality, they inherit the data we give them and the goals we set. That means they can — and often do — learn our biases. Whether it's denying loans to minorities, showing fewer job ads to women, or misidentifying people of color in facial recognition systems, biased AI isn't an edge case. It's a systemic risk.

Bias in AI isn't only a technical flaw. It reflects social values, institutional history, and power dynamics. And because AI systems scale quickly and invisibly, even small biases can have massive consequences. Fairness in AI isn't about perfection — it's about vigilance, transparency, and making intentional choices that serve all communities.

---

1. Where Bias Begins: Data, Design, and Deployment

Bias can enter a system at many points:

Historical Data

AI learns from past patterns. But if the past was unfair, those patterns may encode discrimination.

- Hiring algorithms trained on historical resumes may favor men.
- Criminal risk scores may overestimate risk for minorities.

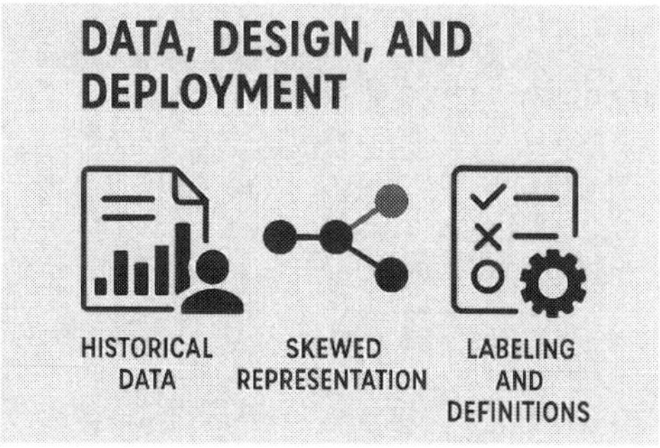

Skewed Representation

If a model is trained mostly on English speakers, urban residents, or Western medical patients, it may perform poorly for rural communities, minority dialects, or global populations.

Labels and Definitions

Training a model to identify "qualified candidates" or "high-risk individuals" depends on how those outcomes are defined. These labels often carry hidden assumptions — such as using arrest history as a proxy for danger or equating higher income with success.

Case Study – Amazon Resume Filter

Amazon built a resume-screening system trained on a decade of hiring history dominated by male applicants. The AI eventually penalized resumes referencing women's organizations. Despite technical sophistication, it simply reflected structural bias — and was scrapped.

Design bias is silent but powerful. It rarely announces itself, yet it shapes who gets access, who gets excluded, and who gets harmed.

---

2. Detecting and Measuring Fairness

Fairness isn't a single metric — different definitions exist, and they often conflict.

Common approaches include:

Demographic Parity

Outcomes should be evenly distributed across groups (e.g., gender or race).

Equal Opportunity

Among qualified candidates, each group should have equal likelihood of success.

Predictive Parity / Calibration

If a model assigns a 70% score, roughly 70% of those individuals should succeed — regardless of demographic.

Choosing a fairness definition is not just technical.

It determines real-world winners and losers. For example, enforcing demographic parity might reduce individual-level fairness to balance group-level outcomes.

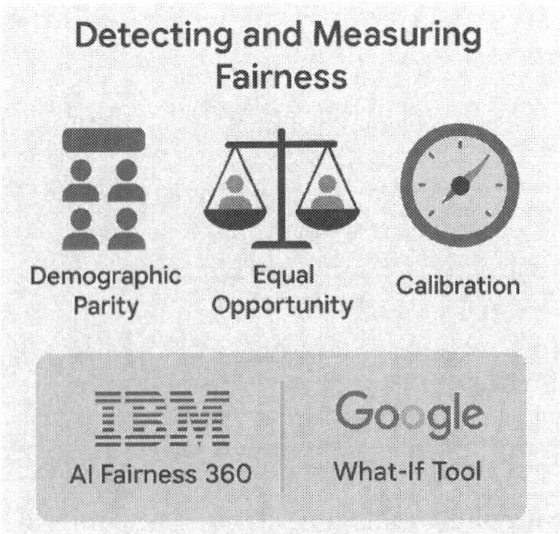

Toolbox Tip

Fairness tools like IBM AI Fairness 360, Google's What-If Tool, and Microsoft's Fairlearn help detect disparities.

But they only work when developers know which fairness criteria are appropriate — and why.

3. Reducing Bias in Practice

Bias mitigation is not a one-time fix. It requires ongoing effort:

Diverse and Representative Training Data

Include voices, contexts, and examples from marginalized or underrepresented groups.

Fairness-Aware Algorithms

Some models incorporate fairness constraints directly into training to balance error rates.

Audits and Benchmarks

Evaluate performance across demographic groups — not just global accuracy.

Multidisciplinary Teams

Engineers alone can't solve fairness. Ethicists, sociologists, domain experts, and community stakeholders must be involved.

Example

Healthcare models often underrepresent Black patients for specific diseases. Rebalancing datasets and involving public health experts leads to more equitable — and clinically more accurate — models.

Fairness isn't about hiding bias. It's about s**urfacing it, naming it, and deciding — collectively — how to address it.**

---

4. Beyond Code: Social and Structural Bias

Not all bias begins in the algorithm. Often, it starts in the world the algorithm is asked to model.

Predictive policing, for example, often directs more patrols to neighborhoods that were historically over-policed — not because of real current crime, but because the data reflects past enforcement patterns.

AI forces us to ask deeper questions:

- Who defines "success"?
- Whose data shapes the model?
- Who benefits — and who is harmed?
- Is AI being used to reinforce control, or expand opportunity?

Ethics is not a technical layer added at the end.

Fairness begins with the values embedded in the system's design.

5. The Way Forward: Building Accountable AI

Fair AI requires structural, not just technical, safeguards:

Transparency

Users and regulators need visibility into how systems work and what data they rely on.

Appeal Mechanisms

People affected by automated decisions should have a way to challenge or contest them.

Impact Assessments

Regular evaluation of real-world outcomes across communities — not just test-set accuracy.

Legal and Regulatory Frameworks

Emerging regulations (e.g., the EU AI Act or algorithmic accountability laws) aim to enforce fairness, transparency, and non-discrimination.

True fairness isn't the absence of bias — it's the presence of accountability, clarity, and shared decision-making.

Quick Takeaways: Bias and Fairness in AI

- **Bias Has Roots in Data**

Bias comes from historical patterns embedded in data and decision-making.

- **Fairness Isn't One-Size-Fits-All**

Fairness has multiple definitions and needs to be chosen with context in mind.

- **Technology Helps, But Isn't Enough**

Technical tools can help—but real fairness requires inclusive design and social awareness.

Chapter Summary

AI doesn't create bias from nothing — it learns it from us. What makes AI risky is not that it's flawed, but that it applies flawed patterns at scale, invisibly and rapidly. But AI can also help detect unfairness, highlight disparities, and improve decisions when designed responsibly.

Creating fair AI requires technical rigor, ethical foresight, and broad participation. The goal isn't perfection — it's transparency, accountability, and continuous improvement. In the end, fairness is a choice we make, not a property the machine magically possesses.

Chapter 25

AI AND JOBS
DISRUPTION, ADAPTATION, AND OPPORTUNITY

Introduction: The Work Shift

Will AI take my job? It's one of the most common questions people ask when thinking about artificial intelligence. The truth is more nuanced than a simple yes or no. AI is transforming work, but not always in the ways we expect. Some tasks are disappearing, others are changing, and entirely new roles are emerging.

People fear job loss because work is tied to identity, stability, and dignity. AI is not simply a technological shift — it's a human one. Understanding how work is changing can help us adapt with confidence rather than anxiety.

This chapter explores the complex relationship between AI and employment. We'll look at which sectors are most affected, what kinds of skills are becoming more valuable, and how humans and AI might collaborate in the workplace of the future.

---

1. Automation vs. Augmentation

Not all job loss is job destruction. Many roles are being reshaped rather than removed. AI is best at automating narrow, repetitive tasks. For example:

- **In finance**, algorithms scan documents and detect fraud.
- **In manufacturing**, robots handle quality checks and assembly.
- **In customer service**, chatbots handle common queries, passing complex ones to humans.

But in each case, AI often augments human workers:

- Fraud teams interpret anomalies flagged by AI.
- Engineers supervise and refine robotic workflows.
- Customer service agents manage emotional or unique cases.

It's important to understand that AI typically automates tasks, not full jobs.

Most occupations include dozens of activities — some repetitive, some interpersonal, some analytical. AI tends to replace the routine parts while amplifying the human ones. The future of work is less about roles disappearing and more about roles being reshaped from the inside.

---

2. Jobs Most at Risk — and Most Resilient

A 2023 McKinsey report estimates that up to 25% of U.S. jobs may be significantly affected by AI by 2030. And while estimates vary across countries, the pattern is similar worldwide: tasks that are repetitive or predictable face the highest pressure, while people-oriented and adaptive roles remain more resilient.

Most vulnerable roles include:

- Data entry clerks
- Paralegals
- Basic accounting
- Warehouse sorting

These roles rely heavily on rule-following, repetition, and structured inputs — areas where AI excels.

More resilient roles include those requiring:

- Emotional intelligence (teachers, therapists, negotiators)
- Physical dexterity and onsite adaptation (plumbers, electricians, mechanics)
- Creativity and strategy (designers, analysts, project managers)

Economists call this "job polarization":

high-skill and low-skill work continues growing, while middle-skill, routine work compresses. AI accelerates that trend by automating predictable tasks but amplifying demand for roles that require judgment, coordination, or hands-on problem-solving.

---

3. New Roles in the AI Economy

While some jobs disappear, new ones are being created:

- AI trainers and prompt engineers
- Model auditors and AI ethicists
- Human–AI collaboration designers
- Data quality analysts

In many cases, existing roles are evolving:

- Marketing managers now use AI to test campaigns.

- HR professionals use AI to screen resumes.
- Journalists use it to generate story drafts.

The economy isn't shrinking — it's shifting.

---

4. Reskilling and the Learning Curve

To stay relevant, many workers will need to adapt. That doesn't always mean learning to code. Often it means learning how to:

- Work with AI tools effectively
- Interpret AI-generated insights
- Think critically about machine output
- Communicate clearly and solve open-ended problems

Governments, companies, and schools all play a role in this transition. Programs that provide digital literacy, upskilling, and lifelong learning will be essential.

Skills Rising in Value

- **AI fluency:** Knowing how to work with and question AI tools.
- **Analytical judgment:** Interpreting insights rather than accepting them blindly.
- **Human-centered skills:** Empathy, communication, leadership, relationship-building.
- **Creative problem-solving:** Framing problems, not just executing tasks.

These skill categories apply across industries and will define who thrives in an AI-driven workforce.

---

5. Ethical and Economic Considerations

The benefits of AI-driven productivity must be balanced with social responsibility:

- If companies automate, do they also reinvest in training?
- How are gains distributed — do they benefit workers or only shareholders?
- How do we prevent a widening gap between AI-augmented and AI-displaced workers?

Policymakers are considering measures such as:

- Job transition support and unemployment benefits

- Tax incentives for companies that retrain workers
- Regulations encouraging ethical AI use in HR and labor practices

The path forward requires coordination between governments, employers, and workers — not just technology.

Quick Takeaways: AI and Jobs

» Automation and Augmentation Coexist

AI doesn't just eliminate jobs — it reshapes them. Routine tasks are automated, while humans handle oversight, complexity, and emotion.

» Jobs Most Vulnerable

Roles that involve repetitive data work or predictable physical tasks — like data entry or warehouse sorting — face the highest risk of automation.

» Human Strengths Still Matter

Jobs that rely on empathy, creativity, adaptability, or nuanced decision-making remain more resilient in the AI era.

» New Careers Are Emerging

Expect rising demand for AI trainers, ethicists, prompt engineers, and others who guide or interpret AI systems.

» Reskilling Is Essential

Workers don't need to become coders — but they do need to understand and collaborate with AI. Lifelong learning will be key.

Chapter Summary

AI is not the end of work — but it is the end of work as we know it. The rise of intelligent tools demands adaptation from workers, employers, and policymakers alike.

Routine tasks are automated. Human skills become more valuable. Entirely new professions emerge. With foresight and fairness, AI can become a force not just for efficiency, but for opportunity.

But that future won't build itself. It requires investment in people, equity in transition, and a shared vision for what work should look like in the age of intelligent machines. The future of work will be human + AI — not human versus AI.

Chapter 26

IMPLEMENTING AI ETHICS PRINCIPLES INTO PRACTICE

Introduction: Ethics by Design, Not as an Afterthought

Talking about AI ethics is not enough. As AI systems become embedded into daily life and critical infrastructure, the conversation must shift from "What are the risks?" to "How do we build ethical systems from the start?"

This chapter explores practical ways to embed ethics into the design, development, and deployment of AI. It's not just about avoiding harm, but about actively aligning AI systems with human values like fairness, accountability, transparency, and dignity. Ethical implementation isn't a final checkpoint — it's a continuous commitment across the entire AI lifecycle.

---

1. Principles First: Building a Foundation

Effective ethical AI starts with a clear set of guiding principles. Common themes adopted by leading companies and institutions include:

- **Fairness:** Avoid discrimination or unjust outcomes across age, gender, race, or socioeconomic status.
- **Transparency:** Make AI decisions and processes explainable and visible to the people affected by them.
- **Privacy and Consent:** Protect personal data and clarify how it's collected, used, and stored.
- **Accountability:** Clearly define who is responsible for decisions and errors — especially when AI informs or automates high-stakes actions.
- **Human-Centered Design:** AI should support—not replace—human autonomy, judgment, and well-being.

Well-established frameworks help teams operationalize these ideas:

- IBM emphasizes "pillars of trust," focusing on fairness, robustness, and transparency.

- Microsoft's Responsible AI Standard highlights human oversight, safety, and continuous auditing.
- Google's AI Principles stress social benefit, privacy, and safety.

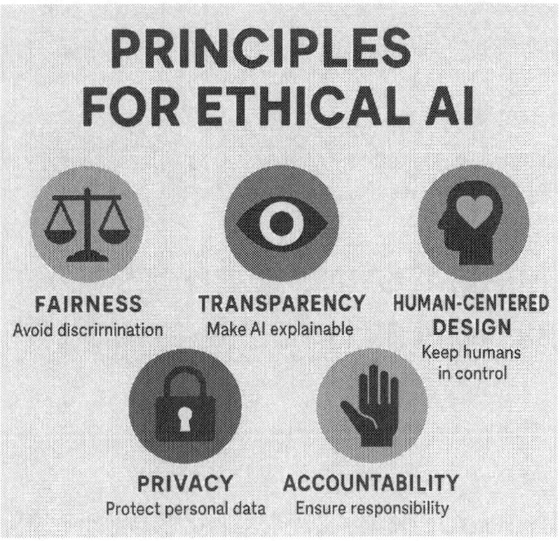

These principles don't solve problems on their own — but they anchor all future choices.

---

2. Design Thinking for Ethics

Ethics isn't a technical patch added at the end. It must be integrated early in the design process.

This includes:

Dichotomy Mapping
Understanding both benefits and potential harms of a feature.

Example: An AI that recommends hotel upgrades may improve customer experience — but can also risk unfair profiling if trained on biased data.

Scenario Planning
Teams ask:

- What if the model gets it wrong?
- Who is affected?
- How visible is the failure?
- Can the user recover from it?

Inclusive Design
Engage diverse stakeholders — including the people most likely to be impacted — early in the process. Homogeneous teams miss real-world perspectives.

A common tool here is the ethical design checklist, where teams ask:

- Does this feature treat users fairly?
- Are training datasets diverse and validated?

- Can we explain how the system makes decisions?
- Would I feel comfortable if this affected someone in my own family?

These questions create accountability before the first line of code is written.

---

3. Guardrails and Governance

Ethical AI requires more than good intentions — it needs structure.

Organizations should implement:

- **Governance Committees:** Groups that review high-risk projects, model changes, or deployments.
- **Guardrails:** Clear rules like "do not sell user data," "always allow human override," or "do not use this system for decisions that affect legal rights without human review."
- **Audit Trails:** Logs of model behavior, data use, and changes over time.
- **Impact Assessments:** Reviews of how an AI system affects different groups, especially vulnerable populations.

Regulation is also evolving.

For example, the **EU AI Act,** introduces strict standards for transparency, documentation, and risk management — particularly in healthcare, education, transportation, and law enforcement.

Governance turns ethical principles into operational reality.

---

4. Tools for Ethical Development

Open-source and commercial tools can bring ethical principles into daily engineering practice:

- **AI Fairness 360 (IBM):** Tests for bias in datasets and model outputs.
- **AI Explainability 360:** Provides methods for interpreting black-box models.
- **FactSheets (IBM):** Standardized "nutrition labels" documenting how a model works, when it performs well, and its limitations.
- **Private AI Approaches:** Running models locally or in secure environments to prevent data leakage and strengthen privacy.

These tools don't remove the need for human oversight, but they make ethical workflows more consistent and repeatable.

---

5. Training, Culture, and Accountability

Technology alone cannot build ethical AI. Culture determines how tools are used and how decisions are made.

Teams should:

- **Train developers and designers** in bias mitigation, responsible data use, and human-centered design.
- **Promote cross-functional collaboration** — involving ethicists, social scientists, legal experts, and end-users alongside engineers.
- **Create safe escalation paths** so team members can raise ethical concerns without fear of retaliation.
- **Reward ethical behavior** — not just speed, performance, or innovation.

Ethics isn't a blocker. It's a compass.

When teams are supported and aligned, ethical thinking accelerates responsible innovation rather than slowing it down.

Quick Takeaways: How to Implement AI Ethics

» Principles First

Start with clear ethical foundations: fairness, transparency, privacy, accountability, and human-centered design.

» Ethical by Design

Use design thinking — like scenario planning and inclusive development — to anticipate and prevent harm.

» Guardrails and Governance

Put systems in place to monitor, audit, and guide AI behavior — before and after deployment.

» Open-Source Tools Help

Resources like IBM's AI Fairness 360 and FactSheets can help detect bias and explain decisions.

» Culture Matters

Ethical AI requires more than code — it needs people trained to care, speak up, and collaborate across disciplines.

Chapter Summary

AI ethics isn't just about avoiding harm — it's about building systems that reflect human values from the very beginning. Ethical AI demands intention, structure, and transparency: through principled design, governance frameworks, technical tools, and a culture that rewards responsibility.

When ethics becomes part of the workflow rather than an afterthought, AI becomes safer, more inclusive, and more trustworthy. The future of responsible AI depends on people who are empowered to design it — thoughtfully, openly, and with care.

Chapter 27

PRIVACY, HALLUCINATIONS, AND THE TRUST PROBLEM

Introduction: When You Can't Tell What's Real

As AI systems become more advanced and embedded in daily life, the boundary between truth and fabrication grows increasingly blurry. Hyper-realistic photos, synthetic voices, and fluent chatbots can generate content that looks, sounds, and feels authentic — even when it isn't grounded in fact. At the same time, these systems often rely on large amounts of personal data to operate, raising serious questions about privacy, consent, and control.

This chapter explores how hallucinations, data use, and synthetic media challenge public trust in AI — and what techniques, policies, and design choices can help rebuild that trust. We introduced hallucinations earlier in the book, but here we examine why they matter so deeply for credibility, safety, and accountability.

---

1. Hallucinations: When AI Makes Things Up

Large language models generate text by predicting the most likely next word, based on patterns from enormous datasets. This makes them excellent at producing coherent, conversational responses — but it also means they may generate statements that sound confident yet are factually incorrect. These errors are known as hallucinations.

Common examples include:

- citing academic papers that don't exist
- inventing statistics or legal precedents
- confidently misrepresenting historical facts
- fabricating steps in a technical process

These aren't "lies" — the system doesn't intend to deceive. It's simply making statistical guesses without the ability to verify truth. It may also have:

- **a fixed knowledge cutoff**
- **limited context windows**, losing track of earlier details
- **no inherent connection** to verified sources

This makes hallucinations particularly risky in healthcare, legal work, hiring, finance, and any domain requiring accuracy.

Reducing Hallucinations — Better Architecture, Not Just Better Prompts

Several approaches help ground an AI model's answers:

- **Structured Prompting:** Clearer instructions reduce ambiguity.
- **Domain-Tuned Systems:** AI assistants trained on a company's own documents or verified internal data perform better within their domain.
- **Fact-Checking Layers:** Some systems include validators that compare outputs against known sources.

But the most important shift is **Retrieval-Augmented Generation (RAG).**

RAG: A More Reliable Way for AI to Answer

RAG systems retrieve relevant documents first, then generate an answer based directly on that information. Instead of guessing from memory, the model summarizes or quotes real content.

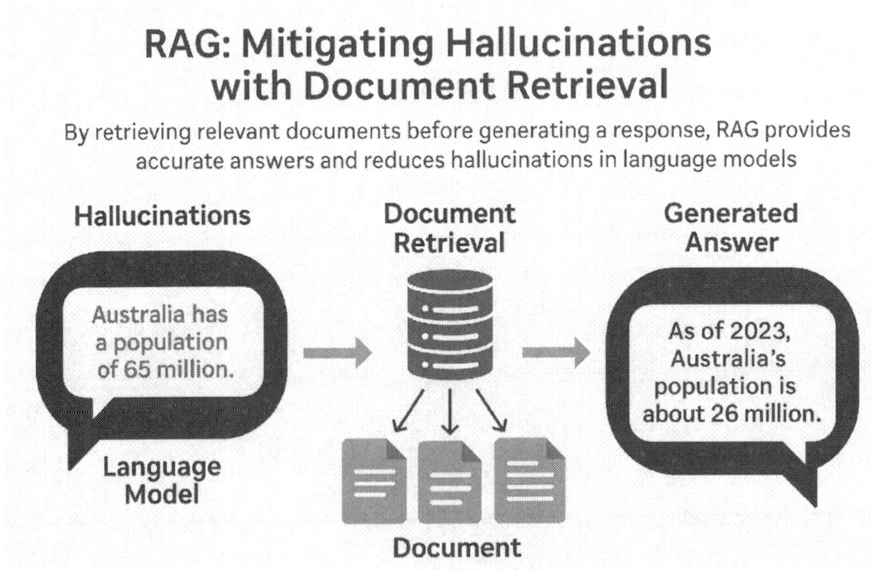

Benefits include:

- **Up-to-date information**, not limited by training cutoffs
- **More accurate**, specific responses
- **Better handling of long, detailed material**
- **Lower likelihood of hallucinations**

RAG isn't perfect, but it represents one of the most promising paths toward trustworthy AI — especially in professional or regulated environments.

---

2. Privacy and Data Collection: The Hidden Cost of AI

Modern AI models are often trained on massive datasets scraped from the internet. This can include:

- social media posts
- blog comments
- online photos
- public forums
- sometimes even emails or documents shared publicly

While this broad training improves language fluency, it creates serious privacy risks:

- People may have no idea their data was used.
- Models can unintentionally reproduce sensitive information.
- Training data may include copyrighted or personal material.

A well-known example is facial recognition systems trained on billions of images scraped without consent — raising debates about ownership, transparency, and privacy rights.

Building Privacy Into AI

Responsible data use requires:

- clear consent policies
- transparent data sourcing
- privacy-preserving techniques such as differential privacy or federated learning
- limits on retention and redistribution

As regulations evolve, privacy protections are becoming a core pillar of trustworthy AI.

3. Deepfakes and Synthetic Media: Seeing Isn't Believing

AI-generated photos, videos, and voice clips are now so lifelike that even experts can struggle to identify fakes. This enables creativity — but also misinformation, fraud, and identity manipulation.

Risks include:

- fabricated political speeches
- impersonation scams
- fake evidence used in personal or legal disputes
- manipulated news clips or events

To counter these threats, researchers and companies are developing:

- **watermarks or metadata** that identify AI-generated content
- **content provenance systems** that trace edit history
- **synthetic media detection tools**
- **public education to strengthen media literacy**

The goal isn't to stop synthetic media — but to ensure authenticity can be verified when it matters.

---

4. The UX of Trust: How Design Shapes Perception

Trust isn't just about the algorithm — it's about the interface.

Research shows:

- When AI recommendations are shown before users form their own judgment (forced display), people are more likely to accept the AI's answer uncritically.
- When users choose when to view AI suggestions (optional display), they make more balanced, thoughtful decisions.

This ties back to concepts introduced earlier in the book about how AI shapes human reasoning.

Good UX for Trustworthy AI Includes:

- **Transparency:** Show limitations, sources, and confidence levels.
- **Control:** Let users decide how and when AI appears.
- **Feedback Mechanisms:** Let users flag issues or correct outputs.
- **Explainability:** Provide meaningful reasons behind recommendations.

Design choices are ethical choices. The interface can encourage either healthy skepticism or unhealthy over-reliance.

Quick Takeaways: Trust and AI

» Hallucinations Happen

Even the best models can generate false information. Verification is essential.

» Privacy Matters

AI systems must respect data rights and minimize unnecessary collection.

» Seeing Isn't Always Believing

Deepfakes and synthetic media challenge digital truth.

» Design Shapes Trust

Interfaces influence how users judge and rely on AI.

» Trust Must Be Earned

Transparency, fairness, and accountability are essential to responsible AI.

Chapter Summary

As AI becomes more powerful, trust becomes more precious. Hallucinations reveal the limits of generative models. Data collection raises urgent privacy questions. Synthetic media blurs the line between truth and fabrication. And interface design can tilt users toward skepticism or blind reliance.

But none of these challenges are unsolvable. With better architectures, privacy-preserving techniques, transparency standards, and thoughtful user experience design, AI can earn — not assume — the trust of the people who rely on it.

In the age of intelligent systems, accuracy, privacy, and accountability aren't optional extras. They are the foundation of trustworthy AI.

Chapter 28

THE FUTURE OF AI — WHERE DO WE GO FROM HERE?

Introduction: From Tools to Partners

As AI becomes more capable, it's also becoming more embedded in the fabric of society. From art and education to government and global health, AI is moving beyond isolated use cases to become a core infrastructure. But where is it heading — and what kind of future are we building with it?

This final chapter explores the key technological, social, and ethical trends shaping the future of AI. It acknowledges the risks, celebrates the possibilities, and reminds us that we — all of us — have a say in what comes next.

---

1. Human–AI Collaboration: The Next Evolution

The future of AI is not about replacement — it's about partnership.

- Co-working with AI: Surgeons using AI during procedures. Writers drafting alongside LLMs. Architects optimizing designs with generative tools.
- Augmentation, not automation: Instead of doing your job for you, AI increasingly enhances your decisions and speeds up your process.
- New roles emerging: Expect jobs like AI interaction designer, prompt strategist, and AI ethicist to grow.

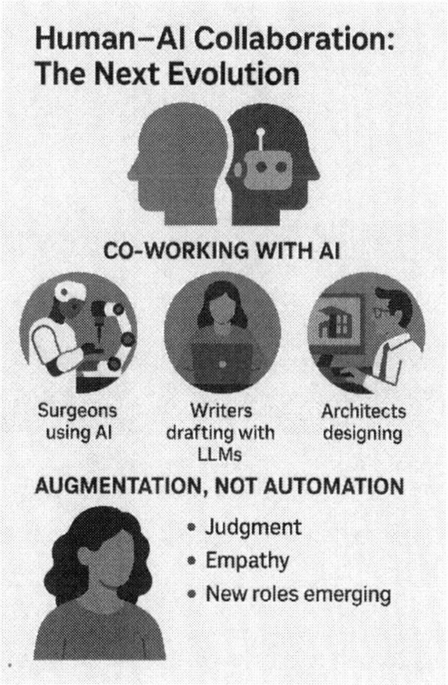

AI will be most powerful when paired with human strengths: judgment, empathy, and creativity.

---

2. The Rise of Multimodal Intelligence

We're entering an era where AI systems no longer process one kind of data at a time.

- **Text, image, voice, video** — combined: Tools like Gemini and GPT-4o can understand a diagram, respond to voice, summarize a page, and describe an image in one interaction.
- **Natural interfaces:** Expect AI to interact via voice, gesture, and AR/VR environments, making tools more accessible across age and ability.
- **Smarter agents:** Multimodal systems will power more context-aware tools — AI that sees, listens, and responds like a partner.

This shift makes AI more useful — but also more complex to govern.

---

3. The Age of AI Agents — From Answers to Actions

A major shift now underway is the rise of **AI agents** — systems that don't just respond to prompts, but take multi-step actions on your behalf.

Unlike chatbots, agents can:

- operate software
- run research workflows
- break tasks into subtasks
- act across apps and websites
- collaborate with other agents
- complete goals end-to-end

Examples include:

- A research agent that searches, analyzes, and summarizes multiple sources.
- A planning agent that books travel by comparing prices and availability.
- A business agent that drafts emails, fills out forms, updates spreadsheets, and monitors deadlines.
- Personal productivity agents that manage tasks automatically.

Agents represent the next evolution of AI:

from assisting with tasks → to completing tasks.

Why this matters

- **Higher autonomy:** Agents can plan, decide, and act in ways traditional assistants cannot.
- **Greater impact on work:** They could automate entire workflows instead of single actions.
- **New risks:** Errors in multi-step decisions multiply quickly; oversight becomes essential.
- **New opportunities:** Individuals and small teams can scale output dramatically.

The challenge ahead

How do we design agents that are powerful — but controllable, transparent, and aligned with human goals?

Agents will reshape not just productivity, but *how we think about delegation, trust, and digital responsibility.*

---

4. Governance, Regulation, and Global Alignment

As AI influences more decisions, it requires oversight.

- **The EU AI Act:** A landmark law categorizing AI risk levels and banning certain uses (like social scoring).
- **NIST AI Risk Management Framework:** A U.S. guideline focusing on trustworthy AI practices.
- **Company standards:** Microsoft, Google, IBM, and others maintain internal review boards and auditing systems.
- **Open vs. closed models:** Debates continue over whether foundational models should be open-source or tightly controlled.

AI governance is not just technical — it's political. Different nations may take different paths, shaping everything from innovation to civil rights.

---

5. Sustainability and the Energy Cost of Intelligence

AI requires computation — and computation uses energy.

- **Training large models:** Systems like GPT or Gemini require significant energy and water for cooling.
- **Inference at scale:** Every user prompt consumes energy, multiplied by millions globally.

Emerging solutions include:

- energy-efficient architectures
- specialized AI chips (e.g., NVIDIA, Apple, AMD)
- renewable-powered data centers

As AI continues to spread, sustainable innovation becomes non-negotiable.

---

6. Societal Shifts and Responsibilities

AI's future is not just about what it can do — it's about what we choose to do with it.

- **AI literacy:** Understanding how AI works will become a foundational skill, like reading or math.
- **Digital divides:** Access to high-quality AI tools will shape opportunity — globally and locally.
- **Global cooperation:** Addressing misinformation, climate change, and labor disruption requires international alignment.
- **Ethical culture:** Developers, educators, businesses, and citizens must all act with transparency and care.

We're not just building tools—we're building norms, expectations, and futures.

Quick Takeaways: The Future of AI

» Human-AI Teams

AI will be a collaborator, not a replacement. Expect tools that work with you, not instead of you.

» Multimodal Capabilities

AI is becoming fluent in language, vision, sound, and more — enabling richer, more natural interactions.

» Global Regulation

Governments and organizations are racing to ensure AI is safe, fair, and accountable.

» Environmental Impact

AI must evolve sustainably — balancing power with planetary responsibility.

» We All Have a Role

The AI future is shaped not just by engineers and CEOs, but by educators, artists, policymakers — and readers like you.

Chapter Summary

AI is not just evolving — it's integrating into how we think, work, and relate. Whether drafting policies, composing content, or automating workflows, the tools of tomorrow will be smarter, faster, and more collaborative. The rise of multimodal systems and autonomous agents will redefine how we interact with technology, bringing both new potential and new responsibility.

But the real future of AI lies not in the code — it lies in the choices we make about how to use it.

Will we use AI to empower, or to exclude? To support, or to surveil?

The answers are not predetermined.

They begin with us.

This concludes Part Four — and the book — with a call not for fear, but for informed engagement. AI's future isn't inevitable. It's intentional.

Conclusion

LIVING WITH INTELLIGENCE

We began this journey with a simple question: **What is artificial intelligence — really — and what does it mean for us?** Along the way, we've unpacked the science, the myths, the breakthroughs, and the risks. We've seen how AI is not just a technology of the future, but a force shaping our present — embedded in our phones, our homes, our workflows, our hospitals, and our creative tools.

But the most important insight isn't about the machines.

It's about us.

AI reflects human values — sometimes clearly, sometimes distortedly. It can amplify creativity, accelerate innovation, and expand access to knowledge. But it can also magnify bias, concentrate power, and erode trust. Which path we follow depends not just on engineers or policymakers, but on all of us — as creators, users, voters, and citizens.

This is no longer a question of whether we live with AI.

We already do.

The real question is: **how?**

- Will we treat AI as a tool that strengthens human judgment — or surrender decisions to black-box systems?
- Will we design with fairness and inclusion — or allow datasets to inherit the blind spots of history?
- Will we automate carelessly — or thoughtfully, with dignity, oversight, and purpose?

There is no final, definitive answer in this book. Instead, there is an invitation: to stay informed, stay curious, and stay engaged. The future of AI is not yet written. It will be shaped by how we choose to use it, question it, regulate it, and integrate it into our lives.

We are not passengers on this journey.

We are co-pilots.

And if we remain thoughtful, responsible, and open-minded, AI can become one of the most powerful allies we have ever known—supporting human potential, expanding opportunity, and helping us build systems that reflect the very best of who we are.

Thank you for taking this journey.

Let's help shape an intelligent future worth living in.

Acknowledgments

This book—and the broader Mastering AI project — would not exist without the people who helped shape it. I am deeply grateful to everyone who contributed their time, insight, and encouragement throughout this journey.

To the readers, learners, and early supporters who believed in the mission before the first chapter was complete — thank you. Your curiosity and your desire to understand AI with clarity and confidence inspired this entire project. Every message, question, and conversation reinforced just how important accessible knowledge has become.

To the researchers, educators, and engineers across the AI community whose work informs this book: your dedication to transparency, scientific rigor, and open dialogue makes responsible AI progress possible. This book reflects and builds upon the foundation you continue to create.

To the Mastering AI team — editors, reviewers, designers, and collaborators — thank you for your commitment to precision, accuracy, and clarity. Your ability to turn complex ideas into accessible knowledge brought this book to life. This is truly a collective effort.

To my family, whose patience and support made countless long days and late nights possible—thank you for grounding and motivating me in every step of the way.

And finally, to every future reader: thank you for choosing to learn, question, and engage with the world of artificial intelligence. Your willingness to understand—not just use—this technology is what will ultimately shape a responsible, human-centered AI future.

— Angel Vale

What You Learned

Over the course of this book, you explored the world of Artificial Intelligence from the inside out — how it works, where it shows up in daily life, and how it's reshaping the systems around us. Here's a clear summary of the key insights you now carry with you:

1. What AI Is — and What It Isn't

You learned that AI isn't magic, consciousness, or a mind. It's pattern recognition at scale — powerful, but fundamentally statistical. AI doesn't "understand" the way humans do, but it can analyze, predict, and generate with remarkable speed.

2. How Today's AI Systems Work

From neural networks to large language models, you now understand how modern AI learns from data, detects patterns, and produces output. You saw how training, inference, bias, and hallucination all arise from the way these systems are built.

3. Where AI Shows Up in Everyday Life

In your phone, your home, your workplace, your car — AI already shapes your routines. You learned how smart assistants, recommendation engines, wearable sensors, and automation tools work behind the scenes to enhance daily tasks.

4. How AI Enhances Creativity and Productivity

Whether writing, designing, managing projects, coding, or brainstorming, AI has become a partner in the creative and professional process. You discovered what AI is good at, where humans still excel, and why human guidance remains essential.

5. The Impact of AI on Work and the Economy

You explored which jobs are changing, which are at risk, and which new roles will emerge. AI disrupts tasks — not entire professions — and the future belongs to workers who understand how to collaborate with intelligent tools.

6. AI in Critical Fields — From Healthcare to Transportation

You saw how AI supports diagnosis, drug discovery, clinical trials, logistics, and even self-driving systems. These aren't science fiction — they're real technologies improving decision-making, safety, and efficiency today.

7. The Risks: Bias, Privacy, Hallucinations, and Misinformation

You learned how AI can unintentionally amplify unfairness, fabricate confident errors, leak private information, or generate deceptive media. Trustworthy AI demands awareness, caution, and good design.

8. How to Think About AI Ethically and Responsibly

Fairness, transparency, accountability, human oversight, and data dignity aren't buzzwords — they're foundations. You learned practical ways to evaluate and implement ethical AI in real-world settings.

9. The Future of AI—and Our Role in Shaping It

AI's trajectory isn't predetermined. You explored future trends—from multimodal intelligence to global regulation — and learned why human values, human judgment, and human intention must guide the next chapter of technical progress.

10. A New Mindset for Living with AI

Above all, you learned that AI should be seen as a tool for empowerment, not replacement. The future belongs to people who stay curious, stay informed, and stay engaged — co-pilots in the journey ahead.

Mastering AI — The Learning Ecosystem

Artificial intelligence is changing the way we live, work, learn, and create.

Mastering AI exists to help people navigate that change with clarity, confidence, and practical skill.

Our mission is simple:

Make AI understandable and useful for everyone.

Whether you're a beginner, a parent, a professional, a student, or simply curious, the Mastering AI ecosystem is designed to support your learning journey.

▸ The Mastering AI Book Series

Understanding AI from every angle

The series covers a wide spectrum of topics so readers can choose what matters most to them. Current and upcoming titles (in English and Spanish editions) include categories such as:

- **AI Essentials** — fundamentals for beginners and everyday users
- **Productivity & Work** — tools, workflows, and smarter ways to get things done
- **Creativity & Content** — writing, design, music, visuals, and storytelling
- **Families & Education** — AI guidance for parents, kids, teens, and teachers
- **Business & Careers** — opportunities, strategy, and navigating workplace change
- **Entrepreneurs & Small Business Owners** — building and scaling with AI
- **Healthcare & Wellness** — diagnostics, monitoring, and responsible health technology
- **Ethics, Governance & The Future** — risks, regulation, fairness, and global impact

New books continue to be added as the technology evolves.

You can explore all current titles at **MasteringAI.ai**.

► **Courses & Workshops**

Practical learning for real-world skills

Mastering AI offers hands-on courses built for different experience levels:

- Foundations of AI
- Practical AI Tools in Action
- Prompting and Problem-Solving
- AI for Business Growth
- The AI Creativity Lab

Courses are designed to be clear, engaging, and focused on real-life applications.

► **Tools, Guides & Tutorials**

Learn faster. Apply instantly.

The ecosystem includes resources created for everyday use:

- Step-by-step How-To Guides
- Short visual tutorials
- Templates and workflows
- YouTube lessons and breakdowns
- Custom GPTs for personal and business needs
- Practical examples you can apply immediately

Everything is designed around accessibility and clarity — whether you're starting from zero or expanding your skill set.

► **Custom AI Assistants**

Built for individuals, creators, and businesses

Mastering AI also develops tailored AI assistants that help with:

- content creation
- customer support
- workflow automation
- research and summarization
- personal productivity

These tools help individuals and teams apply AI in focused, reliable, and safe ways.

▸ Stay Connected

Your journey with AI doesn't end here.

If you'd like updates on new books, guides, courses, or tools, visit:

MasteringAI.ai

You can also join the newsletter for practical AI tips, updates, and new learning resources as they're released.

Together, we can build an intelligent future — one that's accessible, empowering, and designed for everyone.

About the Author

Angel Vale writes as part of the Mastering AI team — a collaborative initiative committed to making artificial intelligence understandable, practical, and empowering for everyone. Bringing together experience in technology, communication, and digital strategy, the team focuses on translating complex concepts into clear, accessible insights that help readers use AI wisely in everyday life.

The *Mastering AI* series spans books, guides, courses, and educational resources designed to support individuals, professionals, and communities as they navigate a rapidly changing technological world. Each title is built with the same mission: to inform, to clarify, and to empower — regardless of background or experience.

This book is one piece of that larger vision: creating an ecosystem where anyone can learn to use AI with confidence, curiosity, and responsibility.

We'd Love Your Feedback!

Thank you for joining us on this journey through the world of Artificial Intelligence.

If this book helped you understand AI more clearly, shifted how you think about technology, or simply sparked your curiosity, your feedback would mean the world to us.

Reviews play a huge role in helping others discover this book—and they help us continue creating clear, accessible AI content for readers everywhere.

How to Leave a Review

If you purchased this book through Amazon or another retailer, you can leave a review on the product page under Customer Reviews.

★★★★★

Just a few words can make a big difference. You might share:

- What part of the book stood out to you
- Whether the explanations helped you understand AI more confidently
- Who you think would benefit from reading it
- If you'd recommend it to others

Your voice matters.

Whether you're an AI beginner, a student, a professional, or a lifelong learner, **your perspective helps shape the conversation around AI literacy and responsible technology.**

Thank you for being part of that conversation—and for supporting thoughtful, accessible writing about AI.

If you'd like more practical tools, guides, and updates from the Mastering AI project, you can visit: **masteringai.ai**

APPENDIX

Glossary of Terms

A

► **Accountability**

Clearly defining who is responsible for an AI system's decisions, behavior, and errors—crucial when automated outcomes affect people's lives.

► **AI (Artificial Intelligence)**

Systems that perform tasks typically requiring human intelligence, such as learning patterns, solving problems, or making predictions.

► **AI Governance**

Policies, oversight structures, and standards designed to ensure AI is used responsibly, safely, and ethically.

► **Algorithm**

A sequence of instructions or rules used to solve a problem or perform a computation.

► **Artificial General Intelligence (AGI)**

A hypothetical AI capable of reasoning, learning, and understanding across any domain — similar to human-level intelligence.

► **Artificial Superintelligence (ASI)**

A theoretical future AI that surpasses human intelligence across all areas (scientific, creative, emotional). Still speculative.

B

► **Bias**

Unfair or skewed behavior in an AI system, often caused by imbalanced or biased training data.

C

► **Calibration**

Ensuring a model's confidence scores reflect real-world accuracy across different groups (e.g., a "70% chance" means 70% in practice).

► **Chatbot**

A system that simulates conversation. Modern chatbots may use large language models (LLMs) to produce natural responses.

► **Context Window**

The amount of text a model can process at once. Larger windows allow the AI to handle longer conversations or documents.

D

► **Data Labeling**

Tagging data — such as classifying images or marking text categories—for use during model training.

► **Data Privacy**

Protection of personal data from misuse, unauthorized access, or unintended disclosure.

► **Deep Learning**

A type of machine learning using multi-layered neural networks to identify complex patterns in data.

► **Demographic Parity**

A fairness metric requiring similar outcomes across demographic groups (e.g., gender or race).

► **Design Thinking**

A human-centered approach to designing AI systems that prioritizes empathy, iteration, and anticipating unintended impacts.

E

► **Ethical AI**

The practice of building and using AI systems aligned with human values such as fairness, privacy, transparency, and accountability.

► **Explainability**

How well an AI system's decisions can be interpreted and understood by humans.

F

► **Fairness**

The effort to avoid discrimination against individuals or groups. Multiple fairness metrics exist depending on context.

G

▸ Generative AI

AI that creates new content — text, images, music, code, or video — based on patterns learned during training.

H

▸ Hallucination

When an AI generates inaccurate or fabricated information that appears credible.

▸ Human-in-the-Loop (HITL)

A workflow where humans supervise, review, or override AI decisions — especially in sensitive or high-risk contexts.

I

▸ Interpretability

Understanding why a model produced a given result — related to explainability but often more technical.

L

▸ Labeling

Assigning categories or attributes to data for training (e.g., "spam" vs "not spam").

▸ Large Language Model (LLM)

An AI trained on massive text datasets to understand and generate human-like language.

M

▸ Machine Learning (ML)

A subset of AI where systems learn patterns from data instead of being manually programmed.

▸ Multimodal AI

AI that can process multiple types of input (e.g., text + images + audio) within a single system.

N

▸ Narrow AI

AI designed for a specific task such as translation, facial recognition, or spam detection.

► **Neural Network**

A mathematical structure inspired by the brain, consisting of interconnected nodes ("neurons") that learn patterns.

P

► **Pattern Recognition**

The ability of AI systems to detect structures, trends, or similarities in data.

► **Prompt**

A user's input or instruction to an AI model.

► **Prompt Engineering**

Crafting precise and effective prompts to guide AI behavior or improve output quality.

R

► **Retrieval-Augmented Generation (RAG)**

A method that retrieves real documents before generating responses, reducing hallucinations and grounding answers in actual information.

T

► **Token**

A unit of text (word or word fragment) used by AI models to measure prompt length and process language.

► **Transformer**

The architecture powering modern LLMs, enabling efficient understanding of long-range relationships in text.

► **Transparency**

Making AI operations visible — how it works, what data it uses, and why it produces specific outputs.

Resources for Further Learning

Whether you're curious to explore AI deeper or want to start building with it, here are curated resources to continue your learning journey.

▸ Books

- **"Artificial Intelligence: A Guide for Thinking Humans"** by Melanie Mitchell

  A thoughtful, accessible look at what AI can—and can't—do.

- **"You Look Like a Thing and I Love You"** by Janelle Shane

  A hilarious and insightful exploration of AI's quirks and limits.

- **"Hello World: Being Human in the Age of Algorithms"** by Hannah Fry

  Explores the social and ethical impact of algorithms in real life.

- **"Life 3.0: Being Human in the Age of Artificial Intelligence"** by Max Tegmark

  A provocative take on AI's future, from near-term to far-reaching consequences.

▸ Online Courses

- **Elements of AI (elementsofai.com)**

  A free, beginner-friendly course co-created by the University of Helsinki and Reaktor.

- **AI for Everyone by Andrew Ng – Coursera**

  A non-technical course explaining AI concepts, use cases, and implications.

- **DeepLearning.AI – ChatGPT Prompt Engineering**
  Learn how to craft better prompts and understand how LLMs work.
- **Google AI and Machine Learning Crash Course**
  Technical and practical for learners with basic coding experience.

▸ AI Tools to Explore

- **ChatGPT** – Natural language assistant for writing, brainstorming, and learning.
- **DALL·E / Midjourney / Adobe Firefly** – Generate images from text prompts.
- **Runway ML** – AI tools for video creation and editing.
- **Udio / Suno / AIVA** – Music generation platforms for all levels.
- **Notion AI / Grammarly** – Enhance writing and productivity with AI support.
- **Khanmigo** – AI tutoring companion integrated with Khan Academy.

▸ Communities and Platforms

- **Hugging Face**
  A hub for open-source AI models, datasets, and research tools.
- **r/MachineLearning (Reddit)**
  A vibrant community for discussions, papers, and practical advice.
- **Partnership on AI**
  Coalition of academics, nonprofits, and companies working on ethical AI.
- **AI Alignment Forum**
  For more technical and philosophical discussions on long-term AI safety.

▸ For Kids and Beginners

- **Teachable Machine by Google**
  A fun way to train simple AI models using your webcam or microphone.

- **Scratch + AI Extensions**

  A way for young learners to play with AI in Scratch programming.

- **Books:**

  - *"AI for Kids"* by Dale Lane
  - *"How to Talk to Robots"* by Tabitha Goldstaub (great for non-tech adults too)

▸ Newsletters & Blogs

- **The Batch** (from Andrew Ng) – Weekly insights into AI trends and research.
- **Import AI** by Jack Clark – Deep dives into AI advances and ethics.
- **MIT Technology Review: The Algorithm** – Digestible summaries of AI news.
- **AI Snake Oil** – Honest, skeptical takes on AI hype and reality.

Made in United States
Orlando, FL
20 January 2026

77098909R00168